Maths Skills

Physics

2nd Edition

Carol Tear

OXFORD
UNIVERSITY PRESS

OXFORD
UNIVERSITY PRESS

Great Clarendon Street, Oxford, OX2 6DP, United Kingdom

Oxford University Press is a department of the University of Oxford.
It furthers the University's objective of excellence in research, scholarship,
and education by publishing worldwide. Oxford is a registered trade mark of
Oxford University Press in the UK and in certain other countries

British Library Cataloguing in Publication Data
Data available

978-0-19-842898-5

10 9 8 7 6 5 4 3 2 1

Printed in India by Multivista Global Pvt. Ltd

Acknowledgements
Cover: Grebesh Kovmaxim/Shutterstock

Artwork by Thomson Digital and Tech-Set Ltd, Gateshead

In-house editor Guy Williamson

Although we have made every effort to trace and contact all
copyright holders before publication this has not been possible in all
cases. If notified, the publisher will rectify any errors or omissions at
the earliest opportunity.

Links to third party websites are provided by Oxford in good faith
and for information only. Oxford disclaims any responsibility for
the materials contained in any third party website referenced in
this work.

Contents

How to use this book   5

1 Measurement   6

1.1   Units and dimensions   6
1.2   Uncertainties and significant figures   8
1.3   Scale of the Universe   10
1.4   A practical activity: The pendulum   12

2 Waves   14

2.1   Graphs of waves   14
2.2   Superposition of waves   16
2.3   Diffraction and interference   18
2.4   Refraction of light   20
2.5   Lenses and stationary waves   22

3 Motion   24

3.1   Motion 1   24
3.2   Motion 2   26
3.3   Motion 3   28
3.4   Forces   30
3.5   Resolving forces   32
3.6   Newton's laws   34
3.7   Work, energy, and power   36
3.8   Vertical motion and gravity   38
3.9   Density, pressure, and upthrust   40
3.10   Momentum   42
3.11   Momentum and energy   44

4 Materials   46

4.1   Elasticity 1   46
4.2   Elasticity 2   48

5 Electricity   50

5.1   Resistance and resistivity   50
5.2   Electric charge and current   52
5.3   emf and potential difference   54
5.4   The potential divider and other circuits   56

6 Quantum physics   58

6.1   The photoelectric effect   58
6.2   Energy, waves, and particles   60

Summary questions for chapters 1-6   62

7 Circular motion and oscillations   64

7.1   Circular motion   64
7.2   SHM 1   66
7.3   SHM 2   68
7.4   Examples of SHM   70

8 Thermal physics   72

8.1   Temperature scales and the gas laws   72
8.2   An ideal gas   74
8.3   Kinetic theory   76
8.4   Internal energy   78

9 Fields   80

9.1   Gravity   80
9.2   Gravitational potential   82
9.3   Orbital motion   84
9.4   Electric fields   86
9.5   Electric potential   88
9.6   Magnetic fields   90
9.7   Electromagnetic induction   92

Summary questions for chapters 7–9   94

10 Capacitors   96

10.1   Capacitors in circuits   96
10.2   RC circuits 1   98
10.3   RC circuits 2   100

11 Radioactivity 102

11.1 Radioactive decay 102

11.2 Activity 104

12 Nuclear physics 106

12.1 Nuclear reactions 106

12.2 Fission and fusion 108

13 Particle physics 110

13.1 Fundamental particles 1 110

13.2 Fundamental particles 2 112

14 Medical physics 114

14.1 X-rays 114

15 Astrophysics 116

15.1 Red shift 116

15.2 The expanding Universe 118

15.3 Stars 120

Summary questions for chapters 10–15 122

Data and formulae 124

Constants and AS formulae 124

A2 formulae 126

Index 128

How to use this book

This workbook has been written to support the development of key mathematics skills required to achieve success in your A Level Science course. It has been devised and written by teachers and the practice questions included reflect the **exam-tested content** for AQA, OCR, and Cambridge specifications.

The workbook is structured into chapters, with each chapter relating to an area of physics. Then, each topic covers a mathematical skill or skills that you may need to practise. Each topic offers the following features:

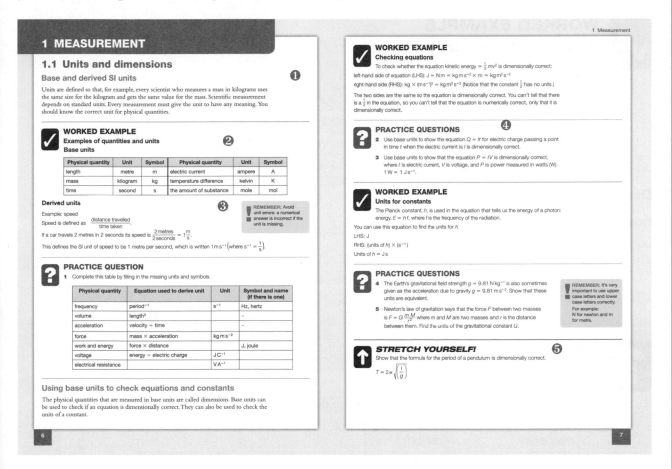

❶ *Opening paragraph* outlines the mathematical skill or skills covered within the spread.

❷ *Worked example* – each spread will have one or two worked examples. The worked examples will be annotated.

❸ *Remember* is a useful box that will offer you tips, hints, and other snippets of useful information.

❹ *Practice questions* are ramped in terms of difficulty and all answers are available at www.oxfordsecondary.co.uk

❺ *Stretch yourself* – some of the spreads may also contain a few more difficult questions and concepts to stretch your mathematical knowledge and understanding.

1.1 Units and dimensions

Base and derived SI units

Units are defined so that, for example, every scientist who measures a mass in kilograms uses the same size for the kilogram and gets the same value for the mass. Scientific measurement depends on standard units. Every measurement must give the unit to have any meaning. You should know the correct unit for physical quantities.

WORKED EXAMPLE

Examples of quantities and units
Base units

Physical quantity	Unit	Symbol	Physical quantity	Unit	Symbol
length	metre	m	electric current	ampere	A
mass	kilogram	kg	temperature difference	kelvin	K
time	second	s	the amount of substance	mole	mol

Derived units

Example: speed

Speed is defined as $\dfrac{\text{distance travelled}}{\text{time taken}}$.

If a car travels 2 metres in 2 seconds its speed is $\dfrac{2\,\text{metres}}{2\,\text{seconds}} = 1\dfrac{\text{m}}{\text{s}}$.

This defines the SI unit of speed to be 1 metre per second, which is written $1\,\text{m s}^{-1}$ $\left(\text{where } \text{s}^{-1} = \dfrac{1}{\text{s}}\right)$.

> **REMEMBER:** Avoid unit errors: a numerical answer is incorrect if the unit is missing.

PRACTICE QUESTION

1 Complete this table by filling in the missing units and symbols.

Physical quantity	Equation used to derive unit	Unit	Symbol and name (if there is one)
frequency	period^{-1}	s^{-1}	Hz, hertz
volume	length^3		–
acceleration	velocity ÷ time		–
force	mass × acceleration	kg m s^{-2}	
work and energy	force × distance		J, joule
voltage	energy ÷ electric charge	J C^{-1}	
electrical resistance		V A^{-1}	

Using base units to check equations and constants

The physical quantities that are measured in base units are called dimensions. Base units can be used to check if an equation is dimensionally correct. They can also be used to check the units of a constant.

WORKED EXAMPLE

Checking equations

To check whether the equation kinetic energy $= \frac{1}{2}mv^2$ is dimensionally correct:

left-hand side of equation (LHS): $J = Nm = kg\,m\,s^{-2} \times m = kg\,m^2\,s^{-2}$

right-hand side (RHS): $kg \times (m\,s^{-1})^2 = kg\,m^2\,s^{-2}$ (Notice that the constant $\frac{1}{2}$ has no units.)

The two sides are the same so the equation is dimensionally correct. You can't tell that there is a $\frac{1}{2}$ in the equation, so you can't tell that the equation is numerically correct, only that it is dimensionally correct.

PRACTICE QUESTIONS

2 Use base units to show the equation $Q = It$ for electric charge passing a point in time t when the electric current is I is dimensionally correct.

3 Use base units to show that the equation $P = IV$ is dimensionally correct, where I is electric current, V is voltage, and P is power measured in watts (W). $1\,W = 1\,J\,s^{-1}$.

WORKED EXAMPLE

Units for constants

The Planck constant, h, is used in the equation that tells us the energy of a photon: energy, $E = hf$, where f is the frequency of the radiation.

You can use this equation to find the units for h.

LHS: J

RHS: (units of h) $\times$ (s^{-1})

Units of $h = J\,s$

PRACTICE QUESTIONS

4 The Earth's gravitational field strength $g = 9.81\,N\,kg^{-1}$ is also sometimes given as the acceleration due to gravity $g = 9.81\,m\,s^{-2}$. Show that these units are equivalent.

5 Newton's law of gravitation says that the force F between two masses is $F = G\frac{mM}{r^2}$ where m and M are two masses and r is the distance between them. Find the units of the gravitational constant G.

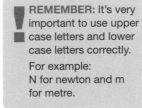

REMEMBER: It's very important to use upper case letters and lower case letters correctly.

For example: N for newton and m for metre.

STRETCH YOURSELF!

Show that the formula for the period of a pendulum is dimensionally correct.

$$T = 2\pi\sqrt{\frac{l}{g}}$$

1.2 Uncertainties and significant figures

Uncertainties

When a physical quantity is measured there will always be a small difference between the measured value and the true value. How important the difference is depends on the size of the measurement and the size of the uncertainty, so it is important to know this information when using data. There are several possible reasons for uncertainty in measurements, including the difficulty of taking the measurement, the precision of the measuring instrument (e.g., owing to the size of the scale divisions), and the natural variation of the quantity being measured. The word uncertainty is generally used in preference to error because the word error is associated with something that is wrong. Mistakes in making measurements should be avoided, not included, in the uncertainty.

✓ WORKED EXAMPLE

Examples of uncertainties in measurements

A measurement of 4.7 g on a scale with divisions of 0.1 g means the value is closer to 4.7 g than 4.6 g or 4.8 g. If the measurement was exactly halfway between 4.7 g and 4.8 g you would round up and record 4.8 g. This means that 4.7 g is anything from 4.65 g up to, but not including, 4.75 g and the measurement is written 4.7 ± 0.05 g.

A length of 6.5 m measured with great care using a 10 m tape measure marked in mm could have an uncertainty of 2 mm and would be recorded as 6.500 ± 0.002 m.

The same length measured with a stick 1 m in length and no scale divisions, in difficult conditions, could have an uncertainty of 0.5 m and would be recorded as 6.5 ± 0.5 m.

It is useful to quote these uncertainties as percentages.

In the first 6.5 m, the percentage uncertainty is $\frac{0.002}{6.500} \times 100\% = 0.03\%$. The measurement is 6.500 m ± 0.03%.

In the second 6.5 m, the percentage uncertainty is $\frac{0.5}{6.5} \times 100\% = 7.69\%$. The measurement is 6.5 m ± 8%.

(Unless the percentage uncertainty is less than 1%, it is acceptable to quote percentage uncertainties to the nearest whole number.)

If the 6.5 m length is measured with a 5% error, the absolute error $= \frac{5}{100} \times 6.5\,\text{m} = \pm 0.325\,\text{m}$.

When a physical quantity is calculated, the uncertainty in the value is equal to the sum of all the percentage errors (not the sum of the absolute errors) in the quantities used in the calculation.

The percentage uncertainty in the area of a rectangle with sides 5.6 ± 0.1 cm and 3.4 ± 0.1 cm is

$\frac{0.1}{5.6} \times 100\% + \frac{0.1}{3.4} \times 100\% = 1.8\% + 2.9\% = 5\%$ (to the nearest whole %).

PRACTICE QUESTIONS

1 Give these measurements with the uncertainty shown as a percentage (to one significant figure):

   a 5.7 ± 0.1 cm   b 2.0 ± 0.1 A   c 450 ± 2 kg

   d 10.60 ± 0.05 s   e 47.5 ± 0.5 mV   f 366 000 ± 1000 J

2 Give these measurements with the error shown as an absolute value:

   a 1200 W ± 10%   b 34.1 m ± 1%

   c 330 000 Ω ± 0.5%   d 0.00800 m ± 1%

3 Identify the measurement with the smallest percentage error.

   A 9 ± 5 mm   B 26 ± 5 mm   C 516 ± 5 mm   D 1400 ± 5 mm

Significant figures

When you use a calculator to work out a numerical answer, you know that this often results in a large number of decimal places and, in most cases, the final few digits are 'not significant.' The uncertainty in the data affects how many figures will be significant. It is important to record your data and your answers to calculations to a reasonable number of significant figures. Too many and your answer is claiming an accuracy that it does not have, too few and you are not showing the precision and care required in scientific analysis.

WORKED EXAMPLE
The number of significant figures
Three significant figures: 2̲7̲1̲ m, 0.2̲7̲1̲ m, 3̲.6̲2̲ m, 0.03̲4̲5̲ m (notice that the zeros here just tell us how large the number is by showing where the decimal point goes. The three significant figures are underlined).

Three significant figures where the zero is significant:

2̲0̲7̲ m (any zero digits between the other significant digits will be significant).

2̲7̲.0̲ m, 0.3̲5̲0̲ m (in these cases extra decimal places are shown as zeros and this means these places are significant; 27 m and 0.35 m have only two significant figures).

Ambiguous significant figures:

270 m (2 or 3?) – This is 2 s.f. unless it is written as 270 m (3 s.f.) or 0.270 km or 2.70×10^2 m (see Topic 1.3).

35 000 kW (2 or more?) – This is 2 s.f. unless it is written as 35 000 kW (3 s.f.) or 35.0 MW or 3.50×10^4 W.

How many significant figures to use?

For practical data, be guided by the uncertainty, as described opposite.

For calculations, use the same number of figures as the data in the question with the lowest number of significant figures. It is not possible for the answer to be more accurate than the data in the question.

PRACTICE QUESTIONS

4 Identify how many significant figures there are in these numbers.

    a 609 W   **b** 3.4 kg   **c** 21.67 m   **d** 400.0 N   **e** 10.01 s

    f 5 MW   **g** 6.0 s   **h** $9.8 \, \mathrm{m \, s^{-2}}$ **i** $3.0 \times 10^8 \, \mathrm{m \, s^{-1}}$

5 Give these measurements to two significant figures:

    a 19.47 m   **b** 115 km   **c** 21.0 s

    d 6.63×10^{-34} J s   **e** 1.673×10^{-27} kg   **f** 5 s

6 Use the equation $V = IR$ to calculate the electric current I through a 3300 Ω resistance R when the voltage $V = 12$ V.

WORKED EXAMPLE
Calculations using uncertainties and significant figures
In the example of the area of a rectangle given opposite, the area = 5.6 cm × 3.4 cm = 19.04 cm².

How many significant figures should you use?

The error was shown to be 5% so the absolute error = $\frac{5}{100} \times 19.04 = 0.952 = \pm 1$ cm².

So the '0' and the '4' are not significant and the answer is between 18 cm² and 20 cm². Two significant figures is appropriate and is the same number as in the two original length measurements. Area = 19 ± 1 cm².

PRACTICE QUESTIONS

State the uncertainty for each of your answers.

7 A car travels 540 m in 16 s. Calculate the average speed.

8 Calculate the circumference and the area of a circular disc with radius 1.4 ± 0.1 cm.

1.3 Scale of the Universe

Distant galaxies

When describing the structure of the Universe you have to use very large numbers. There are billions of galaxies and their average separation is about a million light years. The Big Bang theory says that the Universe began expanding about 14 billion years ago. The Sun formed about 5 billion years ago. These numbers and larger numbers can be expressed in standard form, and by using prefixes.

WORKED EXAMPLE
Using standard form

The diameter of the Earth is $13\,000$ km. $13\,000$ km $= 1.3 \times 10\,000$ km $= 1.3 \times 10^4$ km.

In standard form the number is written with one digit in front of the decimal point and multiplied by the appropriate power of 10.

The distance to the Andromeda galaxy is $2\,200\,000$ light years $= 2.2 \times 1\,000\,000$ ly $= 2.2 \times 10^6$ ly.

PRACTICE QUESTION

1 Give these measurements in standard form:

   a 1350 W   **b** 503 N   **c** 130 000 Pa   **d** 86 400 s

   e 696×10^6 s   **f** 9315×10^5 eV   **g** 0.176×10^{12} C kg^{-1}

WORKED EXAMPLE
Order of magnitude calculations

If a number is rounded to the nearest power of ten you say you are giving an order of magnitude value.

The average separation of the galaxies is $\sim 10^6$ light years. The symbol $\sim$ is used to mean 'to within an order of magnitude.'

The wavelength of red light is 700 nm and that of violet light is 400 nm. They are both a few hundred nanometres so they are the same 'within an order of magnitude'.

PRACTICE QUESTIONS

2 Scientists estimate that the Big Bang occurred 13.7×10^9 years ago. Write this time as an order of magnitude.

3 Identify which planets are the same size to within an order of magnitude.
Radii: Mercury 2.4×10^6 m, Venus 6.09×10^6 m, Earth 6.4×10^6 m, Mars 3.4×10^6 m, Jupiter 7.1×10^7 m, Saturn 6.0×10^7 m, Uranus 2.4×10^7 m, Neptune 2.2×10^7 m.

WORKED EXAMPLE
Prefixes

As an alternative to standard form, these prefixes are used with SI units.
Drax power station has an output of 3.96×10^9 W. This can be written as 3960 MW or 3.96 GW.

> **REMEMBER:** Except for k, the symbols are all upper case. The factors increase in threes, that is: 3, 6, 9, 12.

Prefix	Symbol	Value	Prefix	Symbol	Value
kilo	k	10^3	giga	G	10^9
mega	M	10^6	tera	T	10^{12}

Particle theory

At the other end of the scale, the diameter of an atom is about a tenth of a billionth of a metre. The particles that make up an atomic nucleus are much smaller. These measurements are represented using negative powers of ten and more prefixes.

WORKED EXAMPLE

Powers of ten

One way to understand the negative powers of ten (or any number) is to write out a series and look at the pattern:

$1000 = 10^3$, $100 = 10^2$, $10 = 10^1$, $1 = 10^0$, $0.1 = \frac{1}{10} = 10^{-1}$, $0.01 = \frac{1}{100} = 10^{-2}$, $0.001 = \frac{1}{1000} = 10^{-3}$

To multiply powers of ten, add the indices: $1000 \times 100 = 100\,000$ becomes

$10^3 \times 10^2 = 10^{(3+2)} = 10^5$.

To divide powers of ten, subtract the indices: $\frac{1000}{100} = 10$ becomes

$\frac{10^3}{10^2} = 10^{(3-2)} = 10^1$.

To understand why $10^0 = 1$, think of $\frac{100}{100} = \frac{10^2}{10^2} = 10^{(2-2)} = 10^0 = 1$.

Dividing by 100 (or 10^2) is the same as multiplying by 0.01 (or 10^{-2}).

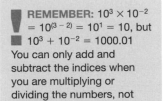

> **REMEMBER:** $10^3 \times 10^{-2}$ $= 10^{(3-2)} = 10^1 = 10$, but $10^3 + 10^{-2} = 1000.01$
> You can only add and subtract the indices when you are multiplying or dividing the numbers, not adding or subtracting them.

PRACTICE QUESTION

4 The speed of light is $3.0 \times 10^8\,\mathrm{m\,s^{-1}}$. Use the equation $v = f\lambda$ to calculate the frequency of:

   a ultraviolet, wavelength $3.0 \times 10^{-7}\,\mathrm{m}$

   b radio waves, wavelength $1000\,\mathrm{m}$

   c X-rays, wavelength $1.0 \times 10^{-10}\,\mathrm{m}$

WORKED EXAMPLE

Small numbers: standard form, orders of magnitude, and prefixes

In standard form the Planck constant $\lambda = 6.63 \times 10^{-34}\,\mathrm{J\,s}$.

The charge on an electron $= 1.6 \times 10^{-19}\,\mathrm{C}$.

As an order of magnitude, the diameter of an atom is $\sim 10^{-10}\,\mathrm{m}$ and that of a nucleus is $\sim 10^{-14}\,\mathrm{m}$.

Prefix	Symbol	Value	Prefix	Symbol	Value
centi	c	10^{-2}	nano	n	10^{-9}
milli	m	10^{-3}	pico	p	10^{-12}
micro	μ	10^{-6}	femto	f	10^{-15}

PRACTICE QUESTIONS

5 Give these measurements in standard form:

   a $0.0025\,\mathrm{m}$ **b** $0.60\,\mathrm{kg}$ **c** $160 \times 10^{-17}\,\mathrm{m}$

   d $0.01 \times 10^{-6}\,\mathrm{J}$ **e** $0.005 \times 10^6\,\mathrm{m}$ **f** $911 \times 10^{-33}\,\mathrm{kg}$

   g $0.00062 \times 10^3\,\mathrm{N}$

6 The charge on an electron is $1.6 \times 10^{-19}\,\mathrm{C}$. Give this as an order of magnitude.

7 Write the measurements for Questions 1a, b, c, f, and g on the previous page and Questions 5a, d, and e above using suitable prefixes.

1.4 A practical activity: The pendulum

Testing a relationship

A graph can be used to test the relationship between two variables and confirm the value of a constant. In this case, it is useful if the relationship between the variables can be arranged as the equation of a straight line. You can then see how closely the plotted points lie on a straight line and this tells you if you have a linear relationship.

✓ **WORKED EXAMPLE**

This practical activity is to show whether the equation relating the period (the time for one swing) of a pendulum to its length is correct. You would need to measure the period, T, for different lengths of pendulum, l, and show that your data is consistent with the equation for the period:

$$T = 2\pi\sqrt{\frac{l}{g}}$$

Where g is the acceleration due to gravity.

If you plot T against l, the graph will be a curve, and you will not be able to tell if the equation is correct. If you change the equation to give a linear relationship then you will be able to tell from the graph if the equation is correct because if you get a straight line the relationship is correct. Squaring both sides of the equation gives:

$$T^2 = 4\pi^2\frac{l}{g}$$

Compare this with the straight line equation: $y = mx + c$.

This equation will give a straight line if you plot T^2 against l, that is: $T^2 = y$, $l = x$

and then $c = 0$ and $m = \dfrac{4\pi^2}{g}$.

So you should get a straight line and, as $c = 0$, this means it should go through the origin. When a straight-line graph goes through the origin, this means that the variables are proportional to each other. This can be written $T^2 \propto l$ (if there is a non-zero value for the constant c the graph doesn't pass through the origin and the relationship is linear, but not proportional).

m is the gradient of the straight line and from this you can calculate a value for g by rearranging the equation for m to give $g = \dfrac{4\pi^2}{m}$. You can compare your value of g with the accepted value of $9.81\,\text{m}\,\text{s}^{-2}$.

To reduce the uncertainty in the data you can time the pendulum for 20 swings. This will be a larger time and so the percentage uncertainty in the measurement will be less.

Graphs of T and T^2 against l for a pendulum

Time for 20 swings/s	T/s	T^2/s^2	l/m
22	1.1	1.21	0.3
26	1.2	1.69	0.4
32	1.6	2.56	0.6
36	1.8	3.24	0.8

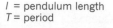

l = pendulum length
T = period

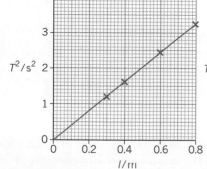

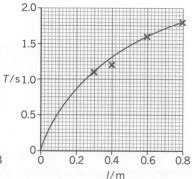

> **REMEMBER:**
> Poor graph plotting is an easy way to lose marks. Make sure you always remember these points:
> 1 Plot the dependent variable against the independent variable, which is the one you are changing.
> 2 Choose scales that use as much as possible of the graph paper. Less than half and you will certainly lose marks.
> 3 Choose scales that are easy to work with. For example, never choose multiples of 3 or 7.
> 4 Label both the axes with the variable separated from its unit by a forward slash. (For example, in the previous figure T^2/s^2. This means that the values of T^2 have units s^2 but you are plotting just a number with no units.)

PRACTICE QUESTION

1 Plot a graph to test the equation using these values of length, l, and time for 20 swings, $t(s)$.

l/cm	20.0	50.0	80.0	100.0	120.0	150.0	175.0	200.0
t/s	18	28	36	40	44	49	53	56

(Hint: The time, t, is for 20 swings, and the units of l are cm. Draw a table and calculate values of T, T^2, and l in m before plotting T^2 against l.)

WORKED EXAMPLE

Using the graph

Looking at the graph of T^2 against l on the previous page, you can say that it is a straight line as it passes through all the points. The line also passes through the origin, showing that $T^2 \propto l$.

To calculate the gradient of the graph:

- Use the largest possible triangle. This makes your value more accurate.

- Use pairs of values that the line passes through, not pairs from the table. It is the gradient of the line you are calculating, so the points must be on it. (If a pair of values is exactly on the line then it makes no difference.)

The gradient of the graph is $m = \dfrac{(3.24 - 0)\,s^2}{(0.8 - 0)\,m} = 4.05\,s^2\,m^{-1}$ (for more about gradients see Topic 3.1).

which gives a value for $g = \dfrac{4\pi^2}{m} = \dfrac{4\pi^2}{4.05\,s^2\,m^{-1}} = 9.74\,m\,s^{-2}$ compared to the true value of $9.81\,m\,s^{-2}$.

If you do this practical activity for yourself you will record the uncertainty in your measurements, perhaps ($\pm0.05\,s$) in the time for one swing and ($\pm0.5\,cm$) in the length. You can then work out a percentage uncertainty for your value of g and calculate whether your value agrees with the true value within the limits of experimental uncertainty.

PRACTICE QUESTION

2 Draw conclusions from your graph and calculate a value for g.

2.1 Graphs of waves

Transverse and longitudinal waves

A wave is produced when a vibrating source periodically disturbs the first particle of a medium. This disturbance is passed from one particle to the next, which creates a wave pattern that travels through the medium. Transverse waves have vibrations at 90° to the direction of travel of the wave. Longitudinal waves have vibrations along the same direction as the wave is travelling.

The frequency, f, at which each individual particle vibrates is equal to the frequency at which the source vibrates and is measured in hertz (Hz). 1 Hz = 1 vibration per second. The period of vibration, T, is measured in seconds and is the time taken for one complete vibration. The period of each individual particle in the medium is equal to the period of vibration of the source.

$$\text{period} = \frac{1}{\text{frequency}}$$

A wave that takes 5 seconds to pass a fixed point has a period of 5 seconds. Its frequency is $\frac{1}{5s} = 0.2\,\text{Hz}$.

A wave that takes 0.1 seconds to pass a fixed point has a period of 0.1 seconds. Its frequency is $\frac{1}{0.1} = 10\,\text{Hz}$. This means that 10 complete waves pass the point in 1 second.

For any wave you can draw a graph to show how far each part of the wave is displaced for different distances along the wave. These are called displacement–distance graphs.

✓ WORKED EXAMPLE

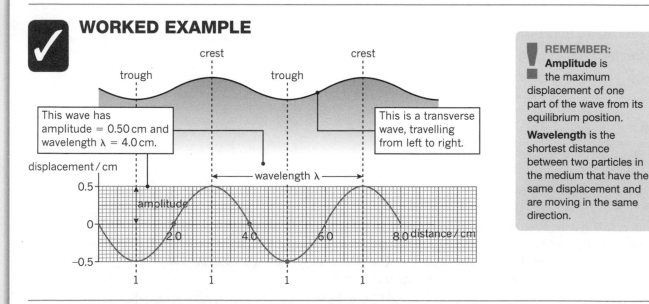

This wave has amplitude = 0.50 cm and wavelength λ = 4.0 cm.

This is a transverse wave, travelling from left to right.

> **REMEMBER:**
> **Amplitude** is the maximum displacement of one part of the wave from its equilibrium position.
>
> **Wavelength** is the shortest distance between two particles in the medium that have the same displacement and are moving in the same direction.

Displacement–time graphs

If you look at a point the wave is passing, you can measure how the displacement changes with time. This is the second type of graph used to show a wave, a displacement–time graph.

WORKED EXAMPLE

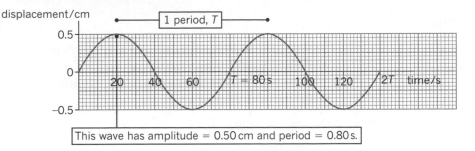

This wave has amplitude = 0.50 cm and period = 0.80 s.

> **REMEMBER:**
> If the data is given to two significant figures, then your answers should not be to more than two significant figures.

PRACTICE QUESTIONS

1 Sketch a graph of the displacement against distance for two complete wavelengths of a wave of amplitude 0.30 metres and wavelength 1.7 metres.

2 Sketch a graph of the displacement against time for two complete oscillations of a wave of amplitude 0.20 metres and time period 3.0 seconds.

The wave equation

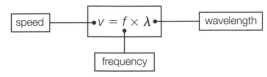

The above equation is known as the wave equation. It states the mathematical relationship between the speed (v) of a wave and its wavelength (λ) and frequency (f).

WORKED EXAMPLE

All electromagnetic waves travel at $3 \times 10^8 \, \text{m s}^{-1}$ in free space. Calculate the frequency of a signal with a wavelength of 1.5 kilometres. The answer can be easily found by dividing both sides of the wave equation by the frequency, λ.
This gives:

$$\frac{v}{\lambda} = f \quad \text{so,} \quad \frac{v}{\lambda} = \frac{(3 \times 10^8 \, \text{m s}^{-1})}{1500 \, \text{m}}$$

$$= 2 \times 10^5 \, \text{Hz}$$
$$= 200 \, \text{kHz}$$

PRACTICE QUESTIONS

3 A sound wave travels through a liquid with a wavelength of 1.5 metres and a frequency of 300 Hz. Calculate the speed of the wave.

4 A sound wave travels through a solid with a wavelength of 2.1 metres and a frequency of 300 Hz. Calculate the speed of the wave.

2.2 Superposition of waves

Phase difference

The phase difference, ϕ, tells us how closely the waves are 'in-step' with each other. If two points, for example, two crests, occur together the waves are 'in phase' or have 'zero phase difference.' If they are *exactly* 'out-of-step', for example, a crest meets a trough, there is a difference of half a wavelength or half a cycle between them and they are 'in antiphase' or 'exactly out of phase'.

Phase difference is measured in degrees (or radians, see Topic 7.1).

Two waves with a phase difference of ϕ

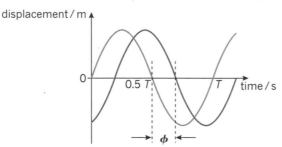

> **REMEMBER:** Phase and phase difference are measured in degrees or radians. It would be incorrect to say 'a phase difference of half a wavelength.'

WORKED EXAMPLE

A wave that starts at the origin is a graph of $y = \sin x$, where x is an angle in degrees and y varies between -1 and $+1$. From $x = 0$ to $x = 360°$ is one complete cycle or period, T, of the wave.

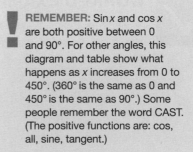

The cosine wave has the same amplitude and period but is shifted along the x-axis by 90° (one quarter of a cycle, $\frac{T}{4}$). The graph is $z = \cos x$.

The phase difference between the sine wave and the cosine wave is 90°.

This means that the graph can also be written as $z = \sin(x + 90°)$.

$x/°$	$y = \sin x$	$z = \cos x$
0	0	1
90	1	0
180	0	-1
270	-1	0
360	0	1
450	1	0

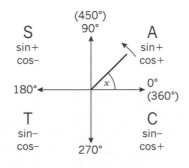

> **REMEMBER:** Sin x and cos x are both positive between 0 and 90°. For other angles, this diagram and table show what happens as x increases from 0 to 450°. (360° is the same as 0 and 450° is the same as 90°.) Some people remember the word CAST. (The positive functions are: cos, all, sine, tangent.)

PRACTICE QUESTION

1 Give the phase difference between waves W and Z in each case.

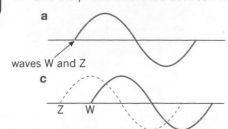

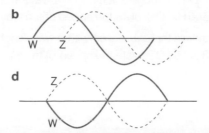

The principle of superposition of waves

When two waves pass through the same point in space and time, the displacement of the wave is the vector sum of the individual displacements.

The displacement is a vector quantity (see Topic 3.1); it is a distance in a certain direction. When displacements are added, the direction is taken into account. For example, equal displacements in opposite directions add to give zero displacement.

Superposition happens whenever waves meet, but the most interesting effects occur when the waves are coherent. Coherent waves have the same frequency and the phase difference between them is constant.

Superposition of two waves, with different amplitude, that are exactly out of phase

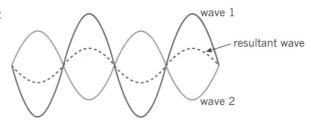

 WORKED EXAMPLE
Superposition and interference

An interference pattern results from the superposition of waves when the waves are coherent. This means they have the same frequency (and wavelength, so they are called monochromatic) and a constant phase difference. As the waves overlap, superposition occurs with constructive interference at some points and destructive interference at other points. Overlapping waves from two coherent sources cause an interference pattern. This can be seen with, for example, water waves, sound waves, and microwaves.

Constructive interference

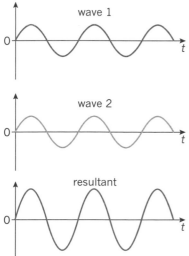

Destructive interference

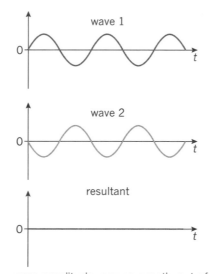

maximum amplitude: waves in phase   zero amplitude: waves exactly out of phase

> **! REMEMBER:** In phase: phase difference = 0, 360° or 2π radians, 720° or 4π radians, −360° or −2π radians, = $n \times 360°$ or $2n\pi$ radians, where n is an integer (a whole number).
>
> Exactly out of phase (in antiphase): phase difference = 180° or π radians, 540° or 3π radians, −180° or −π radians, = $(2n - 1) \times 180°$ or $(2n - 1)\pi$ radians, where n is an integer ($2n - 1$ is always an odd number).

PRACTICE QUESTION
2 **a** For Question 1a opposite, sketch a graph of W and Z and the resultant wave from the superposition of W and Z.

b Repeat for Question 1d.

2.3 Diffraction and interference

Diffraction

When waves pass through a gap of the same order of magnitude as their wavelength they spread out. This is called diffraction. The maximum diffraction occurs when the size of the gap equals the wavelength. This means that two slits can act as two sources of coherent waves; they are coherent because they have come from the same original source.

Interference patterns

With two or more sets of coherent waves there is a possibility that they will later superpose and produce an interference pattern. Laser light from one laser is monochromatic and coherent, and is ideal for producing interference patterns.

WORKED EXAMPLE

Young's slits

Interference occurs because the coherent light waves from the two slits overlap. This results in a series of light and dark fringes on the screen because the distance travelled by the two sets of waves to any point on the screen is different (except for the central point).

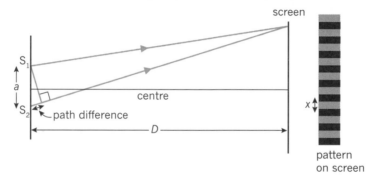

When the path difference is zero (at the central point) or equal to a whole number of wavelengths the waves will be in phase (so the phase difference is 0 or $n \times 360°$ in degrees or $2n\pi$ in radians).

This results in constructive interference and a bright fringe.

When the path difference is equal to an odd number of half wavelengths the waves will be exactly out of phase [so the phase difference is $(2n - 1) \times 180°$ or $(2n - 1)\pi$ radians]. This results in destructive interference and a dark fringe.

The equation for the fringe separation is $x = \dfrac{\lambda D}{a}$ (or sometimes $w = \dfrac{\lambda D}{s}$, so check your formula sheet).

If 10 fringes measure 9.2 mm and the screen is 2.5 m from two slits 1.6 mm apart, calculate the wavelength of the light.

$10x = 9.2\,\text{mm}$ ($x = 9.2 \times 10^{-4}\,\text{m}$), $D = 2.5\,\text{m}$, $a = 1.6\,\text{mm}$

$$9.2 \times 10^{-4}\,\text{m} = \frac{\lambda \times 2.5\,\text{m}}{1.6 \times 10^{-3}\,\text{m}}$$

$$\lambda = \frac{9.2 \times 10^{-4}\,\text{m} \times 1.6 \times 10^{-3}\,\text{m}}{2.5\,\text{m}} = 5.9 \times 10^{-7}\,\text{m (or 590\,nm)}$$

PRACTICE QUESTIONS

1 The light source in the experiment above is changed and 10 fringes measure 8.4 mm. Determine the new wavelength.

2 The experiment in the example above is repeated with light of wavelength:
   a 650 nm   **b** 470 nm.
   Calculate the fringe width in each case.

WORKED EXAMPLE

Diffraction grating

A diffraction grating has many slits, for example, 500 per mm. The waves from all the slits superpose and an interference pattern is produced. In many directions the waves interfere destructively and there are just a few directions where a diffraction pattern is seen.

Diffraction grating

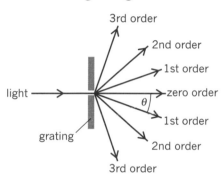

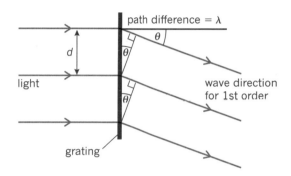

This close-up diagram of part of the grating shows that the 1st order maximum occurs when the path difference between each wave and the next is λ, so for each triangle:

$\sin \theta = \dfrac{\lambda}{d}$ or $d \sin \theta = \lambda$.

The 2nd order maximum will occur when the path difference is 2λ and the nth when it is $n\lambda$.

The general equation for any order is: $d \sin \theta = n\lambda$.

Light with wavelength 650 nm is shone on a grating with 700 lines per mm (2 s.f.). Determine the number of orders seen and calculate the angle of the highest order.

Grating spacing $d = \dfrac{1.0\,mm}{700} = 1.43 \times 10^{-6}\,m$

$n = \dfrac{d \sin \theta}{\lambda} = \dfrac{1.43 \times 10^{-6}\,m \sin \theta}{650 \times 10^{-9}\,m} = 2.20 \sin \theta$

$\sin \theta$ cannot be greater than 1, so maximum n must be 2 orders:

$2 = 2.20 \sin \theta$, $\sin \theta = 0.909$, $\theta = 65°$

PRACTICE QUESTIONS

3 Light is shone on a grating with 550 lines per mm. The 1st order is at an angle of 19°. Calculate the wavelength.

4 Light with wavelength of 710 nm is shone on a grating with 1.0×10^3 lines per mm. Calculate the angle of the 1st order maximum.

5 Light with wavelength 405 nm is shone on a grating with 5.5×10^5 lines per metre. Determine how many orders of lines can be seen.

2.4 Refraction of light

Refractive index

The refractive index, n, of a medium is the ratio of the speed of light in a vacuum to the speed of light in the medium. Light travels fastest in a vacuum so the refractive index is always greater than one ($n > 1$).

The refractive index at a boundary is the ratio of the speed of light in the incident medium to the speed of light in the refracting medium. In this case, the refractive index can be less than one ($n < 1$) because the light may speed up (e.g., when it travels from glass to air).

WORKED EXAMPLE
Refractive index

The refractive index of air, $n_{air} = \dfrac{\text{speed of light in a vacuum}}{\text{speed of light in air}} = \dfrac{c}{c_{air}}$

$$= \frac{2.9979 \times 10^8\,\text{m s}^{-1}}{2.9970 \times 10^8\,\text{m s}^{-1}} = 1.0003$$

1.0003 is 1.00 to 3 s.f., so you usually take the refractive index of air to be the same as that of a vacuum. Notice that as the refractive index is a ratio, it has no units.

The refractive index, n, at the boundary between two different substances with

refractive indices n_1 and n_2 is: $\quad n = \dfrac{\text{speed of light in substance 1}}{\text{speed of light in substance 2}}$

$$= \frac{c_1}{c_2} = \frac{c}{c_2} \times \frac{c_1}{c} = n_2 \times \frac{1}{n_1} \quad \text{so } n = \frac{n_2}{n_1}$$

PRACTICE QUESTIONS

1. The speed of light in water is $2.2540 \times 10^8\,\text{m s}^{-1}$. Calculate the refractive index of water.

2. The refractive index of glass is 1.52. Calculate the speed of light in glass.

3. Use your answer to Question 1 and $n_{glass} = 1.52$ to calculate the refractive index of a water to glass boundary.

WORKED EXAMPLE
Law of refraction

When light changes speed, the wavelength changes and, unless the incident ray is normal to the boundary, the light changes direction.

The law of refraction: $n_1 \sin \theta_1 = n_2 \sin \theta_2$

Refraction of light

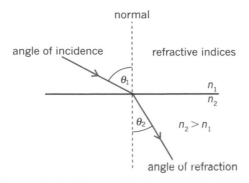

When a light ray passes from water (refractive index 1.33) into glass (refractive index 1.52) at an incident angle of 30°, the angle of refraction is θ where: $1.33 \sin 30° = 1.52 \sin \theta$.

$$\sin \theta = \frac{1.33 \sin 30°}{1.52} = \frac{1.33 \times 0.5}{1.52} = 0.4375, \quad \theta = \sin^{-1} 0.4375 = 25.9° = 26°$$

(Note that $\sin^{-1} 0.4375$ means 'the angle x in degrees for which $\sin x = 0.4375°$.)

PRACTICE QUESTIONS

4 A ray of light in glass ($n_{glass} = 1.52$) is incident on a boundary with ethanol at an angle of incidence of 23°. The angle of refraction is 26°. Calculate the refractive index of ethanol.

5 Light travels from air into garnet. The angle of incidence is 20° and the angle of refraction is 11°. Calculate the refractive index of garnet.

Total internal reflection

When a light ray passes from a dense to a less dense medium (e.g., from glass to air) the angle of refraction is greater than the angle of incidence. In this case, it is possible for the angle of refraction to be equal to, or greater than, 90°. In this situation the ray cannot leave the denser medium and total internal reflection (TIR) occurs.

The critical angle, θ_c, is the angle of incidence that gives an angle of refraction of 90°.

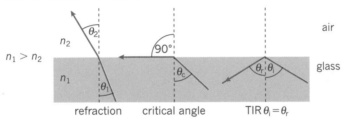

WORKED EXAMPLE

Critical angle

When the angle of incidence is equal to the critical angle, in the diagram above $\theta_1 = \theta_c$ and $\theta_2 = 90°$.

$n_1 \sin \theta_1 = n_2 \sin \theta_2$ and $\sin 90° = 1$, so $n_1 \sin \theta_c = \dfrac{n_2}{n_1}$ is usually written:
$\sin \theta_c = n_2$ where $n_1 > n_2$.

Between glass and air $n_{glass} = 1.5$ and $n_{air} = 1.0$

$\sin \theta_c = \dfrac{1.0}{1.5} = 0.667$, so the critical angle for glass to air $\theta_c = 42°$.

> **REMEMBER:** TIR only occurs when light travels from a dense to a less dense medium. Sin θ_c cannot be greater than 1, so if the equation gives $\sin \theta_c > 1$ you may have the refractive indices the wrong way round – travelling to the denser medium.

PRACTICE QUESTIONS

6 The refractive index of diamond is 2.4. Calculate the critical angle for the diamond–air boundary.

7 The critical angle for the cubic zirconia–air boundary is 27°. Calculate the refractive index of cubic zirconia.

2.5 Lenses and stationary waves

Lenses

A lens is defined by its power, P, measured in dioptres (D), or its focal length, f, measured in metres.

The object distance, u, is the distance of the object from the lens and is positive. The image distance is v and is positive when the image is on the opposite side of the lens to the object. v is negative when the image is on the same side of the lens as the object.

The magnification, m, has no units.

Useful equations are: $P = \dfrac{1}{f}$ $m = \dfrac{v}{u}$ and $\dfrac{1}{f} = \dfrac{1}{u} + \dfrac{1}{v}$ (the lens equation).

Converging lens

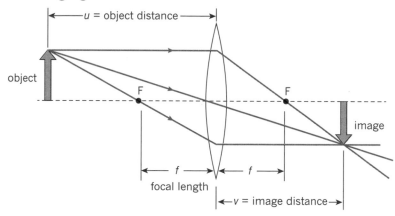

> **REMEMBER:** Converging lenses have positive powers and focal lengths. Diverging lenses have negative powers and focal lengths.
>
> Real images can be formed on a screen. v is positive.
>
> Virtual images are seen through the lens. v is negative.

WORKED EXAMPLE

An object 3 mm high is placed 8 cm from a lens with a power of 5.0 D. Calculate the focal length, image distance, and image size. State if the image is real or virtual.

$$f = \frac{1}{(5.0\,\text{D})} = 0.20\,\text{m}$$

$$(5.0\,\text{D}) = \frac{1}{(8 \times 10^{-2}\,\text{m})} + \frac{1}{v}$$

$$v = \frac{1}{(5.0 - 12.5)}\,\text{m} = -0.13\,\text{m or } 13\,\text{cm} \qquad \text{negative so image is virtual}$$

$$m = \frac{0.13}{0.08} = 1.625 \qquad \text{image size} = 3\,\text{mm} \times 1.625 = 4.9\,\text{mm}$$

PRACTICE QUESTIONS

1   Repeat the calculations above for these lenses:

   a   object 18 mm placed 5.0 cm from a 25 D lens

   b   object 2.0 cm placed 10 cm from a −10 D lens

Stationary waves

Stationary waves, also called standing waves, are produced when two sets of waves with the same amplitude and frequency travelling in opposite directions superpose. There are nodes where the displacement is always zero and antinodes where the displacement varies in time from a maximum positive value, through zero, to a maximum negative value and back again. The distance between adjacent nodes is 0.5λ.

A stretched string

When a string is fixed at both ends and set in vibration it can only vibrate at frequencies that correspond to those wavelengths that allow a node at each end of the string. The lowest possible frequency is called the first harmonic or the fundamental frequency, f_0, and corresponds to the length of the string $L = \frac{1}{2}\lambda$, or $\lambda = 2L$. Higher harmonics correspond to an extra half wavelength fitted into the length of the string, as shown in the diagram.

$v = f\lambda$ and v is constant so as the wavelength doubles the frequency halves.

The 2nd harmonic has wavelength $L = 2 \times \frac{1}{2}\lambda$ so $\lambda = L$ and $f = 2f_0$.

For the nth harmonic $\lambda = \frac{2L}{n}$ and $f = nf_0$.

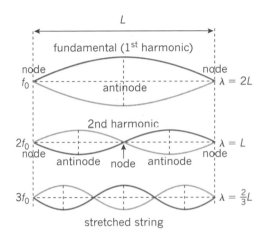

stretched string

Open and closed pipes

When a stationary wave is set up in an open pipe there are antinodes at both ends. The first harmonic, or fundamental, corresponds to half a wavelength fitted in the tube, as shown, and the higher harmonics follow the same pattern as for the string.

When a stationary wave is set up in a closed pipe there is a node at the closed end and an antinode at the open end. The fundamental frequency corresponds to a quarter wavelength as shown. The closed pipe produces only odd harmonics. The nth harmonic will correspond to:

$\lambda = \frac{4L}{(2n-1)}$ and $f = (2n-1)f_0$.

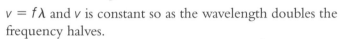

open pipe   closed pipe

WORKED EXAMPLE

A stationary wave is set up on a string 1.2 m long. The first harmonic frequency, f, is 80 Hz. Calculate the wavelength and frequency of the first three harmonics.

$L = 1.2\,m = 0.5\lambda$.

So the wavelength of the fundamental or 1st harmonic $\lambda = 2 \times 1.2\,m = 2.4\,m$ $f = 80\,Hz$

The 2nd harmonic : $\lambda = L = 1.2\,m$ using $v = f\lambda$ as v is constant $f = 2 \times 80\,Hz = 160\,Hz$

The 3rd harmonic: $L = \frac{3}{2}\lambda$ $\lambda = \frac{2L}{n} = \frac{2 \times 1.12\,m}{3} = 0.8\,m$ $f = 3 \times 80\,Hz = 240\,Hz$

PRACTICE QUESTIONS

2   The first harmonic frequency is 147 Hz. List the frequencies and wavelengths of the first four harmonics that can be produced for a 27 cm **a** open pipe, and **b** closed pipe.

3.1 Motion 1

Vectors and scalars

Scalar quantities have magnitude but no direction. Vector quantities have magnitude and direction.

WORKED EXAMPLE
Speed, velocity, and acceleration

Speed is measured in metres per second; per means 'in each'.

You can work out the average speed by measuring the distance travelled and dividing by the time taken. Speed is a scalar quantity; this means it will always be positive. Distance and time are also scalars. They are also always positive.

Velocity is a vector quantity. It is speed in a certain direction. When calculating velocity, v, you need to know the displacement, s. Displacement is the distance travelled in a certain direction. Displacement is a vector. For vectors, the opposite direction is the negative direction. Other directions are sometimes indicated by giving angles.

Acceleration is calculated from the change in velocity, $\triangle v$, divided by the time taken, $\triangle t$. ($\triangle$ is the symbol 'delta' and is used to mean 'a change in'.) Acceleration is the rate of change of velocity, not speed. This means that it is a vector. If an object is moving in a circle at constant speed, its direction is changing. It is accelerating because its velocity is changing.

PRACTICE QUESTIONS

1 State whether each of these terms is a vector quantity or a scalar quantity density, electric charge, electrical resistance, energy, field strength, force, friction, frequency, mass, momentum, power, voltage, volume, weight, work done.

2 For the following data, state whether each is a vector or a scalar: $3\,m\,s^{-1}$, $+20\,m\,s^{-1}$, 100 m NE, 50 km, -5 cm, 10 km S 30° W.

STRETCH YOURSELF!

Work done = force (N) × distance moved in direction of the force (m).
It is measured in joules, 1 J = 1 N m
A moment = force (N) × perpendicular distance from the pivot (m). It is measured in N m.
Explain whether these quantities and units are the same.

WORKED EXAMPLE
Displacement–time graphs

This graph shows the horizontal displacement of the cursor from a point on a monitor screen.

> **REMEMBER:** Always read values to the nearest half, or quarter, of a small square — not to the nearest square.

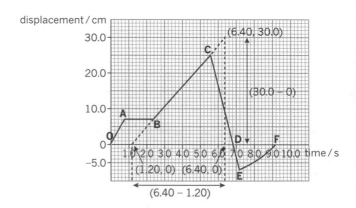

Points to note are: OA constant velocity, AB stationary, BC constant but slower velocity, CE constant, faster, negative velocity. At D the cursor passes through the starting point and continues moving away from it, between E and F the velocity is positive and the cursor is accelerating towards the starting point − the displacement is negative until F when it is zero again.

The velocity between B and C can be found from the gradient of the graph. To calculate the gradient, always draw the largest triangle possible to reduce uncertainties:

$$\text{Gradient} = \frac{(30.0 - 0.0)\,\text{cm}}{(6.40 - 1.20)\,\text{s}} = 5.77\,\text{cm}\,\text{s}^{-1}.$$

Don't forget the units, or to work out a reasonable number of significant figures based on how many you can read from the graph.

Velocity between BC = $5.77\,\text{cm}\,\text{s}^{-1}$ away from the point in the positive direction.

The velocity between E and F is changing. The average speed = the gradient of the straight line EF. The instantaneous speed at a point between E and F is the gradient of the tangent at that point.

WORKED EXAMPLE

Velocity–time graphs

This graph shows how the velocity of an object changes with time.

Points to note from this graph are: the object was stationary at O, accelerated with constant acceleration to P, PQ faster constant acceleration, QR constant velocity, RS slowing down, but not with uniform deceleration, stopping at S. The velocity was always positive so the object was moving in the same direction throughout.

The acceleration between P and Q can be found from the gradient of the graph.

$$\text{Gradient} = \frac{(12.0 - 0.0)\,\text{m s}^{-1}}{(5.20 - 1.00)\,\text{s}} = 2.86\,\text{m s}^{-2}, \quad \text{acceleration} = 2.86\,\text{m s}^{-2}$$

The acceleration between R and S is negative. At any time acceleration is equal to the gradient of the curve at that time. To find the acceleration at $t = 6.0\,\text{s}$, you need to draw a tangent to the curve at the point where $t = 6.0\,\text{s}$ and find the gradient of the tangent. The tangent is the line that 'just touches' the curve at that point.

$$\text{Gradient} = \frac{(0.0 - 12.0)\,\text{m s}^{-1}}{(7.45 - 3.8)\,\text{s}} = -3.29\,\text{m s}^{-2}, \quad \text{acceleration} = -3.29\,\text{m s}^{-2}$$

PRACTICE QUESTIONS

3   For the displacement–time graph on the previous page, calculate the velocity between:
   a   O and A   b   C and E.

4   For the velocity–time graph above, calculate the acceleration:
   a   between O and P   b   at $t = 7.0\,\text{s}$.

3.2 Motion 2

The distance travelled

The distance travelled by an object at constant velocity is equal to the velocity × time of travel. On a velocity–time graph, this is the area of the rectangle between the line and the time axis. For objects travelling with changing velocity, you can imagine that the area can be calculated from lots of very thin rectangles, with widths equal to very small time intervals, and height equal to the average velocity during that time interval.

This leads to the general conclusion that:

> the area under the curve of a velocity–time graph is equal to the distance travelled.

Areas can be calculated using the formulae for the area of a rectangle, triangle, or trapezium, or by 'counting the squares' under a curve.

WORKED EXAMPLE

Triangles and rectangles

The area of the triangle marked A

$$= \tfrac{1}{2} \text{ base} \times \text{height}$$

$$= 0.5 \times 2.40\,\text{s} \times 8.0\,\text{m s}^{-1} = 9.6\,\text{m}$$

The area of the rectangle marked B

$$= \text{base} \times \text{height}$$

$$= (5.20 - 2.40)\,\text{s} \times 8.0\,\text{m s}^{-1} = 22.4\,\text{m}$$

The total distance travelled = 9.6 m + 22.4 = 32.0 m

$$= 32\,\text{m (2 s.f.)}$$

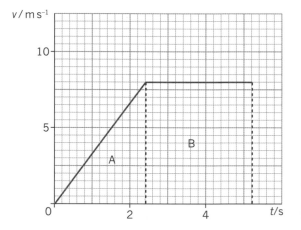

WORKED EXAMPLE

Trapezium

The area of the trapezium

$$= \tfrac{1}{2} \text{ (sum of parallel sides)} \times \text{height}$$

$$= 0.5 \times (7.0 + 16.0)\,\text{s} \times 2.4\,\text{m s}^{-1}$$

$$= 27.6\,\text{m}$$

The total distance travelled = 28 m (2 s.f.)

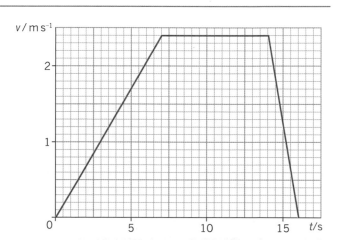

PRACTICE QUESTIONS

1   Use the formula for the area of a trapezium to check the area of the triangles and rectangles worked example above.

2   Use the triangle and rectangle formulae to check the area of the trapezium worked example above.

WORKED EXAMPLE
Counting squares

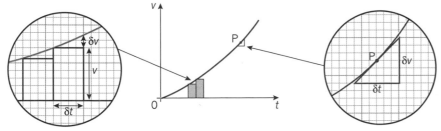

Each small square has an area of $1.0\,\text{s} \times 2.0\,\text{m s}^{-1} = 2.0\,\text{m}$. A large square ($5 \times 5 = 25$) has an area of $25 \times 2.0\,\text{m} = 50\,\text{m}$. It is important to mark off the squares as you count them to avoid missing squares or double counting.

Count all the squares that are half or more under the curve; leave out those where less than half is under the curve.

Total squares = 5 large + 69 small

Distance travelled $\approx 5 \times 50\,\text{m} + 69 \times 2.0\,\text{m} = 250\,\text{m} + 138\,\text{m}$

$$= 388\,\text{m} = 390\,\text{m} \text{ (2 s.f.)}$$

PRACTICE QUESTION

3 Find the distance travelled for the velocity–time graph from Topic 3.1.
(Hint: Use all three methods.)

More about curves

The area under a curve can be found by dividing it into very thin trapeziums of width δt, or sometimes Δt. (The symbol δ is also called 'delta'; it means a very small change.) The total area is the sum of the areas of all the trapeziums. $\sum (v + \frac{1}{2}\delta v)\delta t$ where the symbol $\sum$ means 'the sum of all'. As δt gets smaller and closer to zero (you write this as $\delta t \to 0$), the value for the area gets closer to the true value.

The gradient of the curve at point P can be found by using a triangle where the base δt is very small. The gradient of the straight line at P is $\frac{\delta v}{\delta t}$.

As $\delta t \to 0$ the value for the gradient gets close to the true value.

It is then written $\frac{dv}{dt}$ (called 'd v by d t') and is the gradient of a velocity–time (v–t) graph.

The gradient is the rate of change of v with t and is the acceleration at that point.

PRACTICE QUESTION

4 Determine the value of $a = \frac{dv}{dt}$ at $t = 5.0\,\text{s}$ for the three graphs in the worked examples.

3.3 Motion 3

Kinematic equations

When objects travel with constant acceleration, which can be positive, negative, or zero, their motion is described by a set of equations with these variables:
s = displacement (m), u = initial velocity (ms^{-1}),
v = final velocity (ms^{-1}), a = acceleration (ms^{-2}) and t = time (s).

The equations are: $s = \frac{1}{2}(u + v)t$ $v = u + at$
$s = ut + \frac{1}{2}at^2$ $v^2 = u^2 + 2as$

When you are solving a problem:

- write down the values you know and the ones you want to calculate

- choose the equation and substitute all the values into it

- rearrange the equation and calculate the answer.

 REMEMBER: When acceleration is due to gravity — always use the value of g in the question or from your data sheet.

Gravity acts downwards, which means when an object is moving upwards the acceleration is negative, use $g = -9.81\,ms^{-2}$.

Check which equations are on the formulae sheet.

WORKED EXAMPLES

Amy throws a ball vertically upwards at $5\,ms^{-1}$. When it comes down she catches it at the same point.

a Calculate how high it goes.

Values are: $u = 5.0\,ms^{-1}$, $s = ?$, $v = 0$ (you know this because as it rises it will slow down, until it comes to a stop, and then it will start falling downwards) $a = g = -9.81\,ms^{-2}$

Equation: $v^2 = u^2 + 2as$

Substituting: $(0)^2 = (5.0\,ms^{-1})^2 + 2(-9.81\,ms^{-2})s$

$0 = 25 - 2 \times 9.81 \times s$

Rearranging:

$19.62s = 25$

$s = \dfrac{25}{19.62} = 1.27\,m = 1.3\,m$ (2 s.f.)

b Calculate how long it was in the air.

$u = 5.0\,ms^{-1}$, $s = 0\,m$ (it is back at the starting point) $a = g = -9.81\,ms^{-2}$

Equation: $s = ut + \frac{1}{2}at^2$

Substituting: $(0) = (5.0\,ms^{-1})t + \frac{1}{2}(-9.81\,ms^{-2})t^2$

$0 = t(5.0 - 4.905t)$

(Notice that one solution is at the start, when $t = 0$.)

At the end: $5.0 = 4.905t$

Rearranging: $t = \dfrac{5.0}{4.905} = 1.02\,s = 1.0\,s$ (2 s.f.)

PRACTICE QUESTIONS

1 A car travelling at $13\,ms^{-1}$ accelerates at $4.0\,ms^{-2}$ for $9.0\,s$.
Calculate its final speed.

2 A car travelling at $28\,ms^{-1}$ slows down and stops in $75\,m$.
Calculate the acceleration, assuming it is constant.

3 A stone is dropped down a dry well. It is heard to hit the bottom after $2.9\,s$.
Calculate the depth of the well.

Projectiles

A projectile is a moving object on which the only force acting is gravity. For example, there is no thrust and no air resistance. It could be moving vertically up or down, but more usually it moves both in the vertical *and* the horizontal directions. In this case the path is a parabola. The key to analysing projectile motion is the fact the horizontal and vertical motion can be treated completely separately, using the *suvat* equations on the previous page. In addition when the object is at any point, the time taken to reach that point will be the same in both sets of equations.

WORKED EXAMPLE

A ball is thrown horizontally with a speed of $12.0\,\mathrm{ms^{-1}}$ from a vertical height of $16.0\,\mathrm{m}$. How far has it travelled horizontally when it hits the ground?

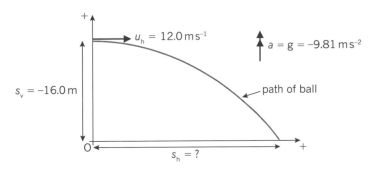

If you find the time it takes to reach the ground in the vertical ($\uparrow$) direction, this will be the same for the travel in the horizontal ($\rightarrow$) direction. Taking up as positive, both s and a will be negative (you could take down as positive but you would need to remember this if you work out the vertical velocity).

$\uparrow$: $u_v = 0$ $s_v = -16.0\,\mathrm{m}$ $a_v = g = -9.81\,\mathrm{ms^{-2}}$ $t = ?$

$$s = ut + \tfrac{1}{2}at^2$$

$$(-16.0\,\mathrm{m}) = \tfrac{1}{2}(-9.81\,\mathrm{ms^{-2}})t^2$$

$$t^2 = \frac{16.0 \times 2}{9.81} = 3.26\,\mathrm{s^2}$$

$$t = 1.81\,\mathrm{s}$$

$\rightarrow$: $u_h = 12.0\,\mathrm{ms^{-1}}$ $a_h = 0$ $t = 1.81\,\mathrm{s}$ $s_h = ?$

$$s = ut + \tfrac{1}{2}at^2$$

$$s_h = (12.0\,\mathrm{ms^{-1}})(1.81\,\mathrm{s}) + \tfrac{1}{2}(0)(1.81\,\mathrm{s})^2$$

$$s_h = 21.7\,\mathrm{m}$$

PRACTICE QUESTIONS

4   A stone is kicked horizontally at $3.0\,\mathrm{ms^{-1}}$ off the top of a cliff $0.2\,\mathrm{km}$ high. Calculate how far from the cliff the stone will hit the beach.

5   A gas molecule has a horizontal velocity of $0.50\,\mathrm{kms^{-1}}$. It crosses a container $0.10\,\mathrm{m}$ wide. Determine the vertical distance the gas molecule falls.

STRETCH YOURSELF!

For Question 4 above, calculate the velocity of the stone when it hits the beach. [Hint: The velocity have a horizontal component and a vertical component and you will need to find the resultant magnitude and the angle (see Topic 3.4).]

3.4 Forces

Resultant forces

Forces are vectors. When vectors are combined their direction must be taken into account. This diagram shows that walking 3 m from A to B and then turning through 30° and walking 2 m to C has the same effect as walking directly from A to C. $\overrightarrow{AC}$ is the resultant vector, denoted by the double arrowhead.

To combine forces, you can draw a similar diagram where the lengths of the sides represent the magnitude of the force (e.g., 30 N and 20 N). The third side of the triangle shows us the magnitude and direction of the resultant force. A careful drawing of a scale diagram allows us to measure these. Notice that if the vectors are combined by drawing them in the opposite order, $\overrightarrow{AD}$ and $\overrightarrow{DC}$, these are the other two sides of the parallelogram and give the same resultant.

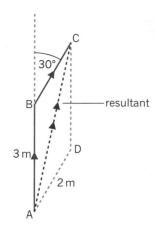

When solving problems you may start by drawing a free body force diagram. The object is a small dot and all the forces acting on it are represented by arrows, which start on the dot.

 WORKED EXAMPLE

A 16 kg mass is suspended from a hook in the ceiling and pulled to one side with a rope, as shown. Sketch a free body force diagram for the mass and a triangle of forces.

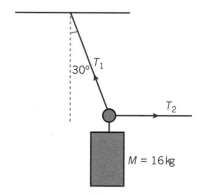

Free body force diagram   **Triangle of forces**

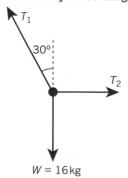   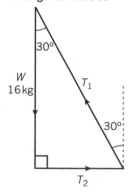

Notice that each force starts from where the previous one ended and they join up to form a triangle with no resultant because the mass is in equilibrium.

PRACTICE QUESTIONS

1 There are three forces on the jib of a tower crane. The tension in the cable T, the weight W, and a third force P acting at X.

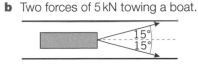

The crane is in equilibrium. Sketch the triangle of forces.

2 For each of these situations draw a triangle or polygon of forces to determine the resultant force:

a

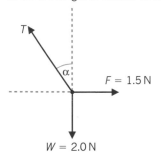

b Two forces of 5 kN towing a boat.

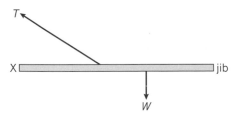

3 These three forces are in equilibrium.
Draw a triangle of forces to find T and α.

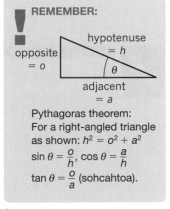

> **REMEMBER:**
>
> hypotenuse = h
> opposite = o
> adjacent = a
>
> Pythagoras theorem:
> For a right-angled triangle as shown: $h^2 = o^2 + a^2$
> $\sin\theta = \dfrac{o}{h}$, $\cos\theta = \dfrac{a}{h}$
> $\tan\theta = \dfrac{o}{a}$ (sohcahtoa).

Calculating resultants

When two forces are acting at right angles the resultant can be calculated using Pythagoras's theorem and the trig functions: sine, cosine, and tangent.

✓ WORKED EXAMPLE

A sub-atomic particle experiences two forces at right angles, one of 2.0×10^{-15} N and the other 3.0×10^{-15} N.

The resultant is represented by F.

$F^2 = (2.0 \times 10^{-15}\,\text{N})^2 + (3.0 \times 10^{-15}\,\text{N})^2$

$F = \sqrt{(4.0 + 9.0)} \times 10^{-15}\,\text{N}$

$F = 3.6 \times 10^{-15}\,\text{N}$

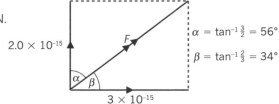

$\alpha = \tan^{-1}\frac{3}{2} = 56°$

$\beta = \tan^{-1}\frac{2}{3} = 34°$

The angle is calculated using either $\tan\alpha$ or $\tan\beta$; remember to state, or show on the diagram, which angle you use. (This diagram shows both, but you only need to calculate one.)

? PRACTICE QUESTION

4 Find the resultant force for these pairs of forces at right angles:

    a 3.0 N and 4.0 N   **b** 5.0 N and 12.0 N

3.5 Resolving forces

Perpendicular forces

Just as you can find the resultant of several forces, you can replace a force with components that have the same effect, when combined, as the original force. This is called resolving a force. Resolving a force into two components at right angles is the key to solving many problems.

This diagram shows that when a force F is resolved into two components at right angles the components are:

$F_1 = F\cos\theta$ and $F_2 = F\sin\theta$ where θ is the angle between F and F_1.

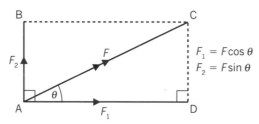

$F_1 = F\cos\theta$
$F_2 = F\sin\theta$

 WORKED EXAMPLE

A block of wood of weight W is placed on a slope. The friction, F, is large enough to prevent it slipping down.

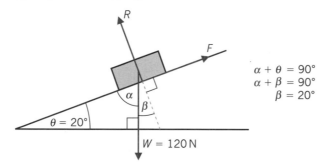

$\alpha + \theta = 90°$
$\alpha + \beta = 90°$
$\beta = 20°$

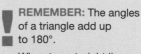
REMEMBER: The angles of a triangle add up to 180°.

When two straight lines intersect, the opposite angles are always equal (because a straight line is 180°).

$\cos(90° - \theta) = \sin\theta$

The forces and motion parallel to the slope are independent from the forces and motion perpendicular to the slope. Resolving in these two directions simplifies the equations as the friction has no effect perpendicular to the slope and the normal reaction, R, has no effect parallel to the slope.

The three forces are in equilibrium so:

Resolving perpendicular to the slope: $R = W\cos\beta = W\cos\theta$

$$R = (120\,\text{N})\cos 20° = 110\,\text{N}$$

Resolving parallel to the slope: $F = W\cos\alpha = W\sin\theta$

$$F = (120\,\text{N})\sin 20° = 41\,\text{N}$$

 PRACTICE QUESTIONS

1 A force of 550 N is applied to a box at an angle of 30° to the horizontal. Calculate the horizontal and vertical components of the force.

2 Calculate the normal reaction and the friction for a box of weight 85 N in equilibrium on a slope of angle 15°.

WORKED EXAMPLE

A lamp hangs from three cables tied as shown.

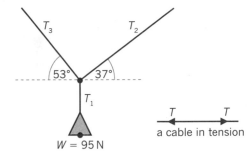

A string or cable can be in tension, but not in compression. If it is in tension, the force pulls on the objects it is attached to. The tension is the same at every point in the cable, or the cable would break, so it is always the same at both ends.

The lamp and the knot in the cable are both in equilibrium.
Draw free body force diagrams for the lamp and the knot in the cable:

For the lamp: $T_1 = 95\,\text{N}$ **(1)**

For the join in the cables: → (resolving horizontally)
$T_3 \cos 53° = T_2 \cos 37°$ **(2)**

↑ (resolving vertically) $T_1 = T_3 \sin 53° + T_2 \sin 37°$ **(3)**

From **(2)**: $T_3 = \dfrac{T_2 \cos 37°}{\cos 53°} = 1.33\,T_2$

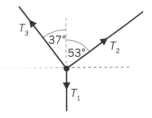

From **(1)** and **(3)** substitute for T_1 and T_3:

$95\,\text{N} = 1.33\,T_2 \sin 53° + T_2 \sin 37°$

$95\,\text{N} = T_2 (1.33 \sin 53° + \sin 37°) = T_2 1.66$

$T_2 = \dfrac{95\,\text{N}}{1.66} = 57\,\text{N}$

From **(2)** $T_3 = 1.33 \times 57\,\text{N} = 76\,\text{N}$

PRACTICE QUESTIONS

3 The three strings in the diagram are in tension and in equilibrium. Calculate the tension in each string.

4 Two masses are supported by three strings. BC is horizontal. Calculate the tension in strings AB, BC, and CD.

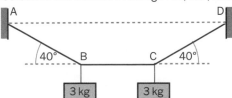

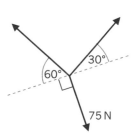

5 A cable, parallel to a slope 30° to the horizontal, pulls a block up the slope at a steady speed. The block weighs 65 N and the friction with the slope is 12 N. Calculate the tension in the cable and the normal reaction force.

3.6 Newton's laws

Newton's first law

When there is no resultant force acting on an object, its velocity stays constant. The object either continues at rest (if its velocity is zero), or it continues to move in a straight line with constant speed (if its velocity is non-zero).

> **REMEMBER:** When an object is moving at constant velocity there is no resultant force on it.

Newton's second law

The rate of change in momentum of an object is directly proportional to the resultant force on the object.

> **REMEMBER:** In the special case where mass is constant the law becomes: acceleration is directly proportional to the resultant force.
>
> Using SI units:
> Resultant force = mass × acceleration
>
> $$F = ma$$

Newton's third law

When an object exerts a force on a second object, the second object simultaneously exerts an equal and opposite force on the first object. (Action and reaction are equal and opposite.)

> **REMEMBER:** Forces always occur as interaction pairs. Don't confuse these with forces on an object in equilibrium. The forces in an interaction pair act on different objects.
>
> The pair of the weight of an object is the gravitational pull from the object on the whole Earth.

WORKED EXAMPLE

A car is pulling a trailer as shown. The driving force is D, friction forces are F_C and F_T, and the tension in the coupling is T.

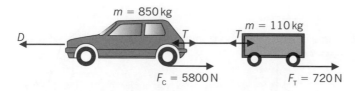

$m = 850\,kg$

$m = 110\,kg$

D T T

$F_C = 5800\,N$ $F_T = 720\,N$

Newton's third law tells us that the force of the coupling pulling on the car, T, is equal and opposite to the force of the car pulling on the coupling, and that the force of the coupling pulling the trailer is equal and opposite to the force of the trailer pulling on the coupling.

1   At constant velocity:

    Newton's first law tells us that the resultant force is zero for both the car and the trailer.

    For the trailer: $\rightarrow T = F_T = 720\,N$

    For the car: $\rightarrow D = T + F_C = 720\,N + 5800\,N = 6520\,N = 6500\,N$ (2 s.f.)

2 Accelerating at $4.0\,\text{m s}^{-2}$ (Newton's second law applies):

Using $F = ma$:

For the trailer: $\to T - F_T = (110\,\text{kg}) \times (4.0\,\text{m s}^{-2})$

$$T = 720\,\text{N} + 440\,\text{N} = 1160\,\text{N} = 1200\,\text{N} \text{ (2 s.f.)}$$

For the car: $\to D - T - F_C = (850\,\text{kg}) \times (4.0\,\text{m s}^{-2})$

$$D = 1160\,\text{N} + 5800\,\text{N} + 3400\,\text{N}$$
$$= 10\,360\,\text{N} = 10\,000\,\text{N} \text{ (2 s.f.)}$$

PRACTICE QUESTIONS

1 A train of mass 740 tonnes accelerates at $0.05\,\text{m s}^{-2}$. Calculate the resultant force on the train. (1 tonne = 1×10^3 kg)

2 A shopper pushes a trolley of mass 28 kg with a horizontal force of 12 N. Calculate the acceleration.

3 An aircraft of mass 1.6×10^4 kg tows a glider of mass 0.6×10^4 kg. When the thrust of the aircraft is 8.3×10^4 N the glider accelerates. Calculate:
   a the acceleration   **b** the tension in the cable

4 Two masses are connected by a string that does not stretch and passes over a frictionless pulley, as shown. Calculate the tension in the string and the acceleration of the masses.

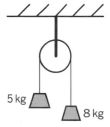

WORKED EXAMPLE

Your weight is $W = mg$, but the sensation of weight comes from the reaction force, R, of the floor pushing on your feet. If you are in a lift that is accelerating, R changes and so you feel as if your weight changes, as shown in these diagrams:

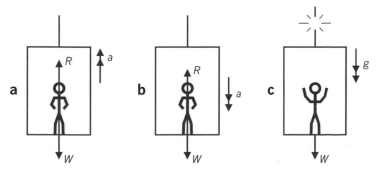

a Going up and speeding up or going down but slowing down:
$R - W = ma$ so $R = W + ma$.

b Going down and speeding up or going up but slowing down:
$W - R = ma$ so $R = W - ma$.

c Cable broken, so accelerating downwards with g:
$W - R = mg$ but $W = mg$ so $R = 0$ so you feel 'weightless'.

3.7 Work, energy, and power

Work done

Work is done when energy is transferred. Work is done when a force makes something move. If work is done *by* an object its energy decreases and if work is done *on* an object its energy increases.

Work done = force × distance moved in the direction of the force.

Work and energy are measured in joules (J) and are scalar quantities (see Topic 3.1).

When force is plotted against the displacement in the direction of the force the area under the graph gives the work done. This gives a way of calculating the work done by a non-uniform force.

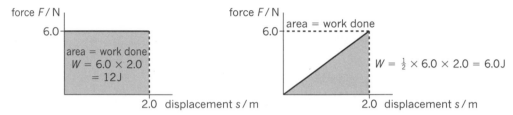

force F/N

6.0

area = work done
$W = 6.0 \times 2.0$
$= 12\,J$

2.0   displacement s/m

force F/N

6.0

area = work done

$W = \frac{1}{2} \times 6.0 \times 2.0 = 6.0\,J$

2.0   displacement s/m

WORKED EXAMPLE

When the force is not in the same direction as the displacement, the distance moved in the direction of the pushing force, F, is $s\cos\theta$.

The work done on the trolley, against the force of friction, by the pushing force is, $W = Fs\cos\theta$.

$W = (25\,N) \times (5\,m)\cos 30° = 108\,J = 110\,J$ (2 s.f.)

The trolley moves at a constant speed and all the energy is transferred as heat (the amount of sound energy is very small).

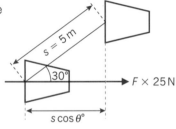

$s = 5\,m$

30°

$F \times 25\,N$

$s\cos\theta°$

WORKED EXAMPLE

The resultant force on a stationary train with mass, m, of 640 tonnes is 380 kN. As there is a net force, the train accelerates. The work done causes an increase in its kinetic energy, ΔE_k. If it travels on a level track for 2.5 km its final speed, v, will be:

$\Delta W = \Delta(\frac{1}{2}mv^2)$ (the symbol delta, Δ, means 'the change in' and $E_k = \frac{1}{2}mv^2$)

$(380 \times 10^3\,N) \times (2.5 \times 10^3\,m) = \frac{1}{2}(640 \times 10^3\,kg)v^2$

$v^2 = \dfrac{2 \times (950 \times 10^6\,J)}{(640 \times 10^3\,kg)} = 2969\,m^2\,s^{-2}$

$v = 54\,m\,s^{-1}$

PRACTICE QUESTIONS

1 Calculate the work done when:

   a the resultant force on a car is 22 kN and it travels 2.0 km

   b a skier of weight 620 N skis for 150 m down a slope at 80° to the vertical.

2 Calculate the kinetic energy gained and the speed of the skier in Question 1b. (Assume the skier starts from rest and frictional forces are small enough to be ignored.)

Power

Power is the rate of work done.

It is measured in watts (W) where 1 watt = 1 joule per second.

$$\text{power} = \frac{\text{energy transferred}}{\text{time taken}} \text{ or power} = \frac{\text{work done}}{\text{time taken}} \quad P = \frac{\Delta W}{\Delta t}$$

WORKED EXAMPLE

A motor lifts a mass, m, of 12 kg through a height, Δh, of 25 m in 6.0 s.

Gravitational potential energy gained:

$$\Delta E_p = mg\Delta h = (12\,\text{kg}) \times (9.81\,\text{m s}^{-2}) \times (25\,\text{m}) = 2943\,\text{J}$$

$$\text{Power} = \frac{(2943\,\text{J})}{(6.0\,\text{s})} = 490\,\text{W (2 s.f.)}$$

WORKED EXAMPLE

A car is travelling at a constant speed of 31 m s^{-1}. The driving force of the engine (to overcome the resistive forces) is 120 N. The power of the engine is:

$$P = \frac{\Delta W}{\Delta t} = \frac{F\Delta s}{\Delta t} \text{ but } \frac{\Delta s}{\Delta t} \text{ is } v \text{ (the velocity) so } P = Fv$$

$$P = (120\,\text{N}) \times (31\,\text{m s}^{-1}) = 3.7\,\text{kW}$$

Efficiency

Whenever work is done, energy is transferred and some energy is transferred to other forms, for example, heat or sound. The efficiency is a measure of how much of the energy is transferred usefully. It is a ratio and is given as a decimal fraction between 0 (all the energy is wasted) and 1 (all the energy is usefully transferred) or as a percentage between 0 and 100%. It is not possible for anything to be 100% efficient: some energy is always lost to the surroundings.

$$\text{Efficiency: } \eta = \frac{\text{useful energy output}}{\text{total energy input}} = \frac{\text{useful power output}}{\text{total power input}}$$

(multiply by 100% for a percentage)

WORKED EXAMPLE

An 850 W microwave oven has a power consumption of 1.2 kW.
Calculate the efficiency.

$$\text{Efficiency} = \frac{\text{useful power output}}{\text{total power input}} = \frac{850\,\text{W}}{1.2 \times 10^3\,\text{W}} = 0.71$$

or as a percentage $0.71 \times 100\% = 71\%$

PRACTICE QUESTIONS

3   A motor rated at 8.0 kW lifts a 2500 N load 15 m in 5.0 s. Calculate the output power and the efficiency.

4   The resultant force on a train is 28 kN and it travels a constant velocity of 45 m s^{-1}.
    a   Determine the useful output power of the train's engines.
    b   If the train's engines are 30% efficient, calculate the input power needed.

5   Determine the time it takes for a 92% efficient 55 W electric motor take to lift a 15 N weight 2.5 m.

3.8 Vertical motion and gravity

Limitations of models

Close to the Earth's surface the acceleration due to gravity is considered to be constant. The value for the acceleration due to gravity, g, is $9.81\,\mathrm{m\,s^{-2}}$. Another way to express this is that the gravitational field strength is $g = 9.81\,\mathrm{N\,kg^{-1}}$. The model: '$g$ is a constant value of $9.81\,\mathrm{N\,kg^{-1}}$' holds well enough close to the Earth for us to use it for many situations.

However, like all models this has some limitations. For example, the value of g is slightly different between the poles and the equator because of the effect of the Earth's rotation and because the Earth is not a perfect sphere. Also, as spacecraft move further away from Earth the effect of the Earth's gravitational field is less, so g can no longer be used.

When analysing the motion of projectiles (see Topic 3.3) you assume there is no air resistance, or drag. The limitations of this model become clear when objects are large, not streamlined or moving at high speed. It is important to know the limitations of the models you use, so that you know whether calculations using the equations are giving sensible answers.

 ## WORKED EXAMPLE

When skydivers fall from an aircraft, and then open parachutes, their acceleration during the fall is not g throughout:

Free-fall from rest

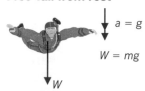

$a = g$

$W = mg$

W

1 Skydivers fall from rest. The resultant force downwards is equal to their weight, $W = mg$. They fall with acceleration g.

Air resistance increases

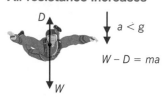

D

$a < g$

$W - D = ma$

W

2 Air resistance, or drag, is proportional to the velocity squared, $D \propto v^2$. As the skydivers speed up the air resistance increases so the resultant force on the skydivers is reduced. This means their acceleration, a, is less than g.

Terminal velocity

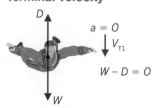

D

$a = 0$

v_{T1}

$W - D = 0$

W

3 When the air resistance equals the weight the resultant force is zero. The acceleration is zero and the skydivers fall with a constant velocity, v_T, called terminal velocity.

Slowing down

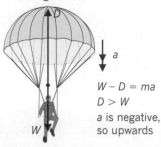

D

a

$W - D = ma$

$D > W$

a is negative, so upwards

W

4 When the skydivers open their parachutes the air resistance (the drag) increases by a large amount. The air resistance is now greater than the weight, $D > W$. The resultant force is upwards so the acceleration, a, is upwards. This means that the skydivers will slow down.

> **REMEMBER:** When the parachute is opened the acceleration is upwards, but this does not mean the skydiver moves upwards. When the velocity is downwards and the acceleration is upwards the skydiver will slow down, but will continue to fall downwards.

Terminal velocity

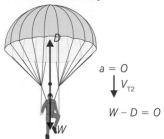

5   As the skydivers slow down the air resistance will decrease and when the air resistance equals the weight they will once again be travelling at a terminal velocity, v_{T2}. This is a much slower terminal velocity than before because the air resistance with the parachute is larger.

✓ WORKED EXAMPLE

An object falling at terminal velocity loses gravitational potential energy but does not gain kinetic energy. The energy heats up the falling object and the surrounding fluid. Spacecraft and asteroids entering the Earth's atmosphere at very high speeds heat up to very high temperatures.

Terminal velocity sketch graph

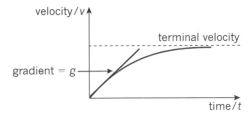

? PRACTICE QUESTIONS

1   A skydiver has a mass, with his parachute, of 77 kg. Calculate the air resistance when he reaches terminal velocity.

2   A skydiver has a mass, with her parachute, of 65 kg. Her terminal velocity is $52\,\mathrm{m\,s^{-1}}$. A model of air resistance is that $D = kv^2$ where D is air resistance in newtons, v is velocity in $\mathrm{m\,s^{-1}}$, and k is a constant. Calculate a value for k.

3   Felix Baumgartner set a record in October 2012 for a speed in free-fall of $370\,\mathrm{m\,s^{-1}}$ and falling from a height of 39 045 m. Explain why he did not reach a lower terminal velocity.

4   Use graph paper to sketch a graph of velocity against time for the skydive shown above using: $v_{T1} = 50\,\mathrm{m\,s^{-1}}$, reached at $t = 27\,\mathrm{s}$, and $v_{T2} = 10\,\mathrm{m\,s^{-1}}$, reached at $t = 46\,\mathrm{s}$.

5   A model rocket is fired vertically upwards using compressed air. The mass of the rocket is 0.30 kg. The initial thrust is 5.0 N. After 2.0 s the thrust is 4.0 N and the air resistance is 0.75 N.
    a   Draw a free body force diagram.
    b   Calculate the initial acceleration.
    c   Calculate the acceleration after 2.0 s.

6   A skydiver, with her parachute, has a mass of 68 kg. She is falling with a velocity of $48\,\mathrm{m\,s^{-1}}$ when she opens her parachute. The air resistance increases to 720 N. Calculate her acceleration.

3.9 Density, pressure, and upthrust

Density

Density, ρ, is the mass per unit volume of an object, or a material. It is measured in $\mathrm{kg\,m^{-3}}$.

$$\text{density} = \frac{\text{mass}}{\text{volume}} \qquad \rho = \frac{m}{V}$$

Most solids and liquids have densities $\sim 10^3\,\mathrm{kg\,m^{-3}}$.

(The symbol $\sim$ means to within an order of magnitude, see Topic 1.3.)

Calculating density often involves calculating the volume of a regular shape.
Check which formulae are given on your formulae sheet and which you need to learn.

REMEMBER: Water has a density of $1\,\mathrm{g\,cm^{-3}}$. This unit is a convenient size but is 1000 times bigger than the SI unit, which is $1\,\mathrm{kg\,m^{-3}}$.

$1000\,\mathrm{g} = 1\,\mathrm{kg}$ but $1\,000\,000\,\mathrm{cm^3} = 1\,\mathrm{m^3}$

Always change densities to $\mathrm{kg\,m^{-3}}$ in calculations to avoid errors. Water has a density of $1000\,\mathrm{kg\,m^{-3}}$.

REMEMBER: Formulae for shapes:

Volumes have dimensions of length cubed. They are measured in $\mathrm{m^3}$ and there will be three lengths in the formula: examples are a cube l^3, cylinder $\pi r^2 l$, and sphere $\frac{4}{3}\pi r^3$.

Areas have dimensions of length squared. They are measured in $\mathrm{m^2}$ and there will be two lengths in the formula: examples are a square l^2, circle πr^2, and surface area of a sphere $4\pi r^2$.

Perimeters have dimensions of length. They are measured in m and there will be one length in the formula: an example is the circumference of a circle $2\pi r$.

WORKED EXAMPLE

The mass of 20 steel ball-bearings is $42\,\mathrm{g}$. Their mean diameter is $8.0\,\mathrm{mm}$.

The volume of one ball-bearing $= \frac{4}{3}\pi r^3 = \frac{4}{3}\pi\,(4.0 \times 10^{-3}\,\mathrm{m})^3$

$\qquad = 2.68 \times 10^{-7}\,\mathrm{m^3}$

The mass of one ball-bearing $= 42 \times 10^{-3}\,\mathrm{kg} \div 20 = 2.1 \times 10^{-3}\,\mathrm{kg}$

density, $\rho = \dfrac{m}{V} = \dfrac{(2.1 \times 10^{-3}\,\mathrm{kg})}{(2.68 \times 10^{-7}\,\mathrm{m^3})} = 7.8 \times 10^3\,\mathrm{kg\,m^{-3}}$

WORKED EXAMPLE

A spherical bubble of air, radius $2.0\,\mathrm{mm}$, under water has volume

$V = \frac{4}{3}\pi r^3 = \frac{4}{3}\pi(2.0 \times 10^{-3}\,\mathrm{m})^3 = 3.35 \times 10^{-8}\,\mathrm{m^3}$

The density of the air $= 1.2\,\mathrm{kg\,m^{-3}}$.

The density of water $= 1.0 \times 10^3\,\mathrm{kg\,m^{-3}}$.

The mass of the bubble of air $m = \rho V = (1.2\,\mathrm{kg\,m^{-3}}) \times (3.35 \times 10^{-8}\,\mathrm{m^3})$

$\qquad = 4.02 \times 10^{-8}\,\mathrm{kg}$.

The weight of the bubble of air $W = mg = (4.02 \times 10^{-8}\,\mathrm{kg}) \times (9.81\,\mathrm{m\,s^{-2}})$

$\qquad = 3.9 \times 10^{-7}\,\mathrm{N}\ (2\ \mathrm{s.f.}).$

PRACTICE QUESTIONS

1 Calculate the density of a solid cube with sides of 5.0 cm and mass of 1.31 kg.

2 Calculate the mass of 150 m of copper cable of circular cross section with diameter 0.50 mm.
(The density of copper is 8.9 g cm^{-3}.)

3 Calculate the volume of mercury in a container if the total mass is 5.5 kg and an empty container has mass 0.45 kg.
(The density of mercury is 13.6×10^3 kg m^{-3}.)

Pressure

Pressure, p, is the force per unit area. It is measured in pascals, Pa, where $1 \text{ Pa} = 1 \text{ N m}^{-2}$.

$$\text{pressure} = \frac{\text{force}}{\text{area}} \qquad p = \frac{F}{A}$$

WORKED EXAMPLE

Air pressure is about 100 kPa. On 1 cm^2 this is a force of:

$F = pA = (1 \times 10^5 \text{ N}) \times (0.01 \text{ m} \times 0.01 \text{ m}) = 10 \text{ N}$

PRACTICE QUESTIONS

4 Calculate the pressure exerted at the point of a pin when it is pushed against a board with a force of 2.0 N, if the area of the point is 0.10 mm^2.

5 A rectangular block with dimensions 6.0 cm × 8.0 cm × 12.0 cm is made of aluminium with a density of 2700 kg m^{-3}.
Find the maximum pressure it can exert when placed on one of its faces on a horizontal surface.

Upthrust

When an object of mass m_o is immersed in a fluid it displaces a volume of the fluid equal to its own volume. There is a force upwards on the object called the upthrust.

Upthrust = weight of fluid displaced. $U = m_f g$ where m_f is the mass of fluid displaced.

If the upthrust is greater than the weight of the object it will accelerate upwards. If the upthrust is less than the weight of the object it will accelerate downwards. Air resistance, or drag, R, may cause the object to move with terminal velocity.

WORKED EXAMPLE

In the worked example at the end of the previous page the bubble of air has weight $Wa = 3.9 \times 10^{-7}$ kg.

The same volume of water has mass $= m_w = \rho_w V = (1.0 \times 10^3 \text{ kg m}^{-3}) \times (3.35 \times 10^{-8} \text{ m}^3)$

$$= 3.35 \times 10^{-5} \text{ kg}.$$

The weight of the water $W_w = m_w g = (3.35 \times 10^{-5} \text{ kg}) \times (9.81 \text{ ms}^{-2}) = 3.3 \times 10^{-4}$ N.

The upward force on the air bubble $= U - W_a = 3.3 \times 10^{-4}$ N $- 3.9 \times 10^{-7}$ N $= 3.3 - 10^{-4}$ N (the weight of the air is negligible).

PRACTICE QUESTION

6 An iron ball of mass 160 g is suspended from a wire and is totally submerged in a liquid of density 810 kg m^{-3}. Calculate the tension in the wire (density of iron = 8.0×10^3 kg m^{-3}).

3.10 Momentum

Impulse and momentum

The momentum of an object p, where m is the mass of the object and v is its velocity, is given by:

$p = mv$

Newton's second law (Topic 3.6) is: force = rate of change of momentum,

$$F = \frac{\Delta p}{\Delta t} \text{ or } F = \frac{\Delta(mv)}{\Delta t}$$

From this you get the relationship that the impulse, $F\Delta t$, is equal to the change in momentum:

$F\Delta t = \Delta(mv)$

Momentum can be measured in kg m s^{-1} or in N s – the two units are equivalent.

Momentum is a vector. It has the same direction as the velocity. When solving problems, it is important to decide which direction is positive and state which values of momentum are negative.

WORKED EXAMPLE

A jet engine travels forward at velocity $195\,\text{m s}^{-1}$ by emitting 75 kg of exhaust gases per second. Calculate the forward force on the jet engine.

Newton's third law: Forward force on rocket = force back on the exhaust gases, F.

In 1.0 s the force on the gases changes the momentum of 75 kg of the gas, which has an initial velocity of $0\,\text{m s}^{-1}$ and a final velocity of $195\,\text{m s}^{-1}$.

$$F = \frac{\Delta(mv)}{\Delta t} = \frac{mv - mu}{\Delta t} = \frac{75\,\text{kg} \times 195\,\text{m s}^{-1} - 75\,\text{kg} \times 0\,\text{m s}^{-1}}{1.0\,\text{s}} = 15\,\text{kN}$$

PRACTICE QUESTIONS

1 A truck of mass 2 tonnes travels at a speed of $30\,\text{m s}^{-1}$. Calculate its momentum.

2 Water from a hose flows at $2.0\,\text{kg s}^{-1}$ and strikes a wall horizontally at a velocity of $3.0\,\text{m s}^{-1}$. Calculate the force on the wall, assuming the water stops moving horizontally after hitting the wall.

3 An object of mass 0.50 kg is travelling at $6.0\,\text{m s}^{-1}$. Calculate the impulse on the object if the final velocity is:

    a $+12\,\text{m s}^{-1}$ b $-12\,\text{m s}^{-1}$

Conservation of momentum

When objects collide, if no external forces act on the objects, the total momentum is the same before and after the collision.

The diagram shows two objects of mass m_1 and m_2 moving with initial velocities before collision of u_1 and u_2, and final velocities after collision of v_1 and v_2.

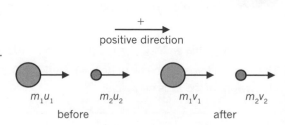

The positive direction is towards the right.

The total momentum before the collision is equal to the total momentum after the collision:

$m_1u_1 + m_2u_2 = m_1v_1 + m_2v_2$

WORKED EXAMPLE

Two objects collide. The car is not moving until the lorry hits it.

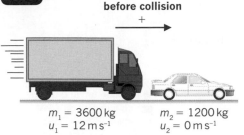

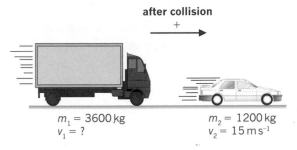

before collision   **after collision**

$m_1 = 3600\,\text{kg}$
$u_1 = 12\,\text{m s}^{-1}$

$m_2 = 1200\,\text{kg}$
$u_2 = 0\,\text{m s}^{-1}$

$m_1 = 3600\,\text{kg}$
$v_1 = ?$

$m_2 = 1200\,\text{kg}$
$v_2 = 15\,\text{m s}^{-1}$

The arrow shows you have defined the positive direction to the right.

Total momentum before $= (3600\,\text{kg}) \times (12\,\text{m s}^{-1}) + (0) = 43\,200\,\text{kg m s}^{-1}$

Total momentum after $= (3600\,\text{kg})v_1 + (1200\,\text{kg}) \times (15\,\text{m s}^{-1})$

By conservation of momentum: $(3600\,\text{kg})v_1 = 43\,200\,\text{kg m s}^{-1} - 18\,000\,\text{kg m s}^{-1} = 25\,200\,\text{kg m s}^{-1}$

so, $v_1 = 7.0\,\text{m s}^{-1}$

WORKED EXAMPLE

Two objects collide and stick together.
$v_1 = v_2$ so $(m_1v_1 + m_2v_2)$ becomes $(m_1 + m_2)v_2$.

The arrow shows you have defined the positive direction to the right.

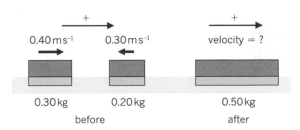

$0.40\,\text{m s}^{-1}$ $0.30\,\text{m s}^{-1}$ velocity = ?

$0.30\,\text{kg}$ $0.20\,\text{kg}$ $0.50\,\text{kg}$

before   after

Total momentum before $= (0.30\,\text{kg}) \times (0.40\,\text{m s}^{-1}) + (0.20\,\text{kg}) \times (-0.30\,\text{m s}^{-1})$
$= (0.120 - 0.060)\,\text{kg m s}^{-1} = 0.060\,\text{kg m s}^{-1}$

Total momentum after $= (0.50\,\text{kg})v_2$

By conservation of momentum: $(0.50\,\text{kg})v_2 = 0.060\,\text{kg m s}^{-1}$ $v_2 = 0.12\,\text{m s}^{-1}$

WORKED EXAMPLE

Two objects explode apart

A firework travels vertically and reaches maximum height. It explodes into two pieces. One piece has a mass of 250 g and moves with a speed of $8.0\,\text{m s}^{-1}$. The other piece has a mass of 120 g. Calculate its velocity.

Total momentum before $= 0\,\text{kg m s}^{-1}$ (because velocity is zero)

Total momentum after $= (0.250\,\text{g}) \times (8.0\,\text{m s}^{-1}) + (0.120\,\text{kg})\,v_2$

By conservation of momentum: $(0.120\,\text{kg})v_2 = -2.0\,\text{kg m s}^{-1}$

$v_2 = -17\,\text{m s}^{-1}$ The negative sign shows this velocity is in the opposite direction to that of the 250 g piece.

PRACTICE QUESTIONS

4 A toy car of mass 0.80 kg and velocity $1.5\,\text{m s}^{-1}$ collides head-on with a car of mass 1.2 kg travelling at $1.8\,\text{m s}^{-1}$ in the opposite direction. They stick together. Calculate their combined velocity after the collision.

5 Two spaceships dock (join together) in space. Spaceship A has a mass of 2500 kg and velocity of $5.0\,\text{m s}^{-1}$. Spaceship B has a mass of 1100 kg and approaches from behind A, travelling in the same direction at $6.0\,\text{m s}^{-1}$. Calculate their combined velocity after docking.

3.11 Momentum and energy

Elastic and inelastic collisions

In an elastic collision all the kinetic energy is conserved. Real collisions are not completely elastic in the macroscopic world, as energy is always transferred to the surroundings, but elastic collisions do occur between atomic and subatomic particles.

In an inelastic collision kinetic energy is not conserved. In a perfectly inelastic collision the two objects stick together; this doesn't always mean that all the kinetic energy is lost, the objects may move with the same velocity.

For all collisions (elastic and inelastic):

- When no external force acts on a system of colliding objects their total momentum is conserved.

- The total energy is always conserved although it may be transferred from one form to another.

> **REMEMBER:**
> Momentum and energy are two very different physical quantities. Energy is a scalar quantity measured in J. Momentum is a vector quantity measured in Ns (or $kg\,m\,s^{-1}$). In problems, momentum and energy must be considered separately.

 WORKED EXAMPLE

A neutron collides head-on with a stationary nucleus of mass twice that of the neutron. The initial velocity of the neutron is $1.2 \times 10^7\,m\,s^{-1}$. After the collision it rebounds with velocity of $0.40 \times 10^7\,m\,s^{-1}$.

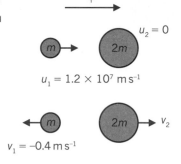

The velocity of the other nucleus is found using conservation of momentum:

$m(1.2 \times 10^7\,m\,s^{-1}) + 0 = m(-0.40 \times 10^7\,m\,s^{-1}) + 2mv_2$

$2v_2 = (1.2 \times 10^7\,m\,s^{-1}) + (0.40 \times 10^7\,m\,s^{-1})$

$v_2 = \frac{1}{2}(1.6 \times 10^7\,m\,s^{-1}) = 0.80 \times 10^7\,m\,s^{-1}$

If the collision is elastic the kinetic energy, E_K, is conserved:

Before: $E_K = \frac{1}{2}m(1.2 \times 10^7\,m\,s^{-1})^2$

After: $E_K = \frac{1}{2}m(0.40 \times 10^7\,m\,s^{-1})^2 + \frac{1}{2}(2m(0.80 \times 10^7\,m\,s^{-1})^2)$

$E_K = \frac{1}{2}m(0.4^2 + 2 \times 0.8^2)(10^7\,m\,s^{-1})^2$

$E_K = \frac{1}{2}m(0.16 + 2 \times 0.64)(10^7\,m\,s^{-1})^2 = \frac{1}{2}m(1.2 \times 10^7\,m\,s^{-1})^2$

so, the collision is elastic.

 PRACTICE QUESTIONS

1 Two snooker balls, A and B, with the same mass move towards each other and collide. The initial velocities are A = $+0.3\,m\,s^{-1}$ and B = $-0.2\,m\,s^{-1}$. The final velocity of A = $-0.2\,m\,s^{-1}$.

   a Find the final velocity of B.

   b Show whether the collision is elastic.

2 A mass of $5.0\,kg$ moving with a velocity of $20.0\,m\,s^{-1}$ to the right collides with a stationary mass of $10.0\,kg$. The final velocity of the $5.0\,kg$ mass is $6.67\,m\,s^{-1}$ to the left.

   a Calculate the final velocity of the $10.0\,kg$ mass.

   b State whether the collision is elastic.

3 An alpha particle of mass 4.0 u with a velocity of $1.0 \times 10^6 \, \mathrm{ms}^{-1}$ to the right collides with a stationary proton of mass 1.0 u. After the collision the alpha particle moves with a velocity of $0.60 \times 10^6 \, \mathrm{ms}^{-1}$ to the right.

    a Calculate the velocity of the proton.

    b Show that the collision is elastic. [1 u = 1 a.m.u. = 1.661×10^{-27} kg, see Topic 12.1] (Hint: for this question it is easier to work in u rather than convert to kg.)

WORKED EXAMPLE

A bullet of mass 5.0 g is fired horizontally into a block of wood of mass 1.0 kg, which is hanging from a long string. The block swings and reaches a height of 6.0 cm. Calculate the speed of the bullet.

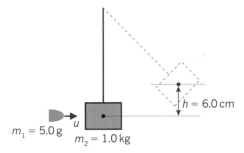

When the bullet hits the wood it will be embedded in the wood and the combined mass will move with velocity v. For the collision of the bullet with the wood, by conservation of momentum:

$(5.0 \times 10^{-3} \, \mathrm{kg})u + 0 = (1.005 \, \mathrm{kg})v$ (equation 1)

For the swing of the pendulum to the highest point, gain in E_P = loss in E_K.

$(\triangle mgh = \triangle \frac{1}{2}mv^2)$

$(1.005 \, \mathrm{kg})(9.81 \, \mathrm{ms}^{-2})(0.06 \, \mathrm{m}) = \frac{1}{2}(1.005 \, \mathrm{kg})v^2$

$v^2 = 2 \times (9.81 \, \mathrm{ms}^{-2})(0.06 \, \mathrm{m})$

$v = 1.08 \, \mathrm{ms}^{-1}$

Substituting this in equation 1, $u = \dfrac{(1.005 \, \mathrm{kg})(1.08 \, \mathrm{ms}^{-1})}{5.0 \times 10^{-3} \, \mathrm{kg}} = 217 \, \mathrm{ms}^{-1}$

$= 220 \, \mathrm{ms}^{-1}$ (2 s.f.)

PRACTICE QUESTIONS

4 A bullet of mass 7.0 g is fired horizontally with velocity of $210 \, \mathrm{ms}^{-1}$ into a wooden block. The block is suspended by a string. Calculate the maximum height the block rises as it swings.

5 A 1.0 kg mass with initial velocity of $5.0 \, \mathrm{ms}^{-1}$ collides with, and sticks to, a stationary 6.0 kg mass. The combined mass collides with, and sticks to, a stationary 3.0 kg mass. The collisions are all head-on.

Calculate:

    a the final velocity

    b the kinetic energy lost.

6 A child of mass 42 kg jumps onto a platform of mass 12 kg suspended from a long rope. The platform swings so that the centre of mass rises 1.6 m. Calculate the initial velocity of the child arriving on the platform.

4.1 Elasticity 1

Hooke's law

A solid may be deformed by stretching it. In this case it is in tension, the forces applied are tensile forces and there is an extension, that is, an increase in length.

A solid may be deformed by compressing it. In this case it is in compression, the forces applied are compressive forces, and there is a decrease in length.

If a material obeys Hooke's law then the extension, x, is directly proportional to the force applied, F, as long as the Hooke's law limit is not exceeded.

$F \propto x$ or $F = kx$ where k is the force constant for the wire (or spring constant for a spring).

$k = \dfrac{F}{x}$ Units of k are $\mathrm{N\,m^{-1}}$ (Note: the extension, x, is sometimes written as a change in length, $\triangle L$, where the original length of the wire is L.)

A metal that obeys Hooke's law

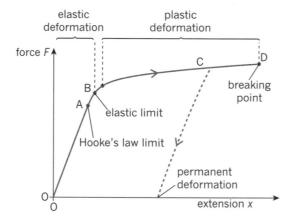

This graph is for a metal. OA is the region where Hooke's law is obeyed. Other metals, and other materials, will have different graphs depending on their properties. For example, a brittle material has no plastic region.

Notice that force is directly proportional to extension. It is incorrect to say that force is directly proportional to the length, L, of the wire. A graph of F against x goes through the origin $(0, 0)$, but on a graph of F against L, when $F = 0$, $L =$ the unstretched length of the wire, L_0.

This is a linear relationship: $F = k(L - L_0)$.

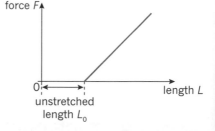

WORKED EXAMPLE

This graph is for a wire stretched with applied force, F, and the extension is x.

The force constant, k, is the gradient of the graph:

$k = \dfrac{56\,\mathrm{N}}{0.75 \times 10^{-3}\,\mathrm{m}} = 75\,000\,\mathrm{N\,m^{-1}}$

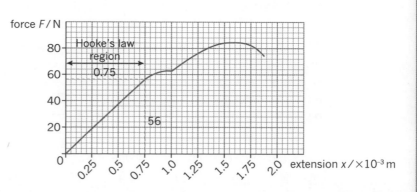

PRACTICE QUESTIONS

1 This data is for the length of a spring.

Force/N	0.0	1.0	2.0	3.0	4.0	5.0	6.0	7.0
Length/cm	2.6	2.8	3.0	3.2	3.4	3.6	3.9	4.3

   a Plot a force–extension graph for the spring.

   b Explain the shape of the graph.

   c Calculate the spring constant.

2 A force of 160 N extends a copper wire by 2.7 cm. Calculate the force constant.

3 In the worked example at the end of the previous page, calculate the breaking force of the wire.

Elastic potential energy

When a force extends a material, work is done by the force. If the material is elastic it then has elastic potential energy, E, which is released when it returns to its original shape. When plastic deformation occurs the force deforms the material permanently and energy is transferred as heat.

If the force was constant whilst the material was stretched, the work done could be calculated from $W = Fx$ (see Topic 3.7). On a force–extension graph this is the area of the rectangle between the line and the extension axis. Although the force is changing, the area under the curve of a force–extension graph is still equal to the work done in extending the material (see Topic 3.7).

In the Hooke's law region, the area can be calculated using the formula for the area of a triangle.

The elastic potential energy, E:

$E = \frac{1}{2}Fx$ and (because $F = kx$) $E = \frac{1}{2}kx^2$

Outside of the Hooke's law region, the work done can be found by 'counting the squares' under a curve (see Topic 3.2).

WORKED EXAMPLE

This is the force–extension graph for a material that stretches obeying Hooke's law but doesn't go back to its original length.

The work done in stretching = area of triangle (A + B)

$E_L = \frac{1}{2} (26\,N) \times (0.34 \times 10^{-3}\,m) = 4.42 \times 10^{-3}\,J$

    $= 4.4 \times 10^{-3}\,J$ (2 s.f.)

The energy recovered when the force is removed

= area of triangle B.

$E_U = \frac{1}{2} (26\,N) \times [(0.34 - 0.1) \times 10^{-3}\,m] = 2.73 \times 10^{-3}\,J = 2.7 \times 10^{-3}\,J$ (2 s.f.)

The energy transferred as heat during stretching = $E_L - E_U$

                                 $= 4.42 \times 10^{-3}\,J - 2.73 \times 10^{-3}\,J$

                                 $= 1.7 \times 10^{-3}\,J$ (2 s.f.)

PRACTICE QUESTIONS

4 For the wire in the example on the previous page:

   a Calculate the energy stored in extending the wire by 0.75 mm

   b Calculate the total energy transferred up to the point where the wire breaks.

4.2 Elasticity 2

Stress and strain

The diagram shows a wire of cross-sectional area, A, under tension from a force F.
The stress, σ, is defined by:

$$\text{stress} = \frac{\text{force}}{\text{area}} \quad \sigma = \frac{F}{A} \quad \text{The unit of stress is the pascal: } 1\,\text{Pa} = 1\,\text{N m}^{-2}$$

The length of the wire L increases by x, the extension of the wire. The strain, ε, is defined by:

$$\text{strain} = \frac{\text{extension}}{\text{original length}} \quad \varepsilon = \frac{x}{L} \quad \text{Strain is a ratio of two lengths, so it has no units.}$$

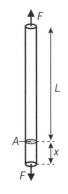

Note that as the wire is in tension, these are tensile stress and tensile strain. There is also compressive stress and strain and shear (twisting) stress and strain.

Stress–strain graphs look similar to force–extension graphs but the advantage is that they are the same for any size and shape of wire made from the same material.

The gradient of a stress–strain graph is the Young modulus of the material:

$$\text{Young modulus} = \frac{\text{tensile stress}}{\text{tensile strain}} = \frac{\sigma}{\varepsilon} = \frac{F \div A}{x \div L} = \frac{FL}{xA} \quad \text{The units are pascals, the same as stress.}$$

 WORKED EXAMPLE

These are stress–strain graphs for a brittle, a strong, a ductile, and a plastic material.
$1\,\text{GPa} = 10^9\,\text{Pa}$

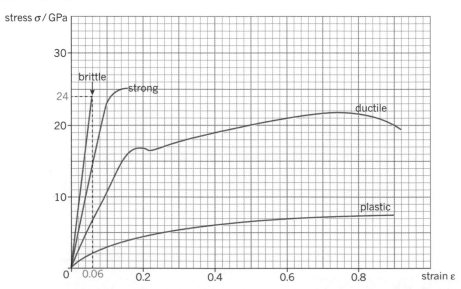

The ultimate tensile strength for the brittle material is $2.40 \times 10^{10}\,\text{Pa}$ (there is no plastic deformation).

The Young modulus for the brittle material = the gradient of Hooke's law region.

$$\text{Young modulus} = \frac{(24.0 \times 10^9\,\text{Pa})}{(0.060)} = 4.0 \times 10^{11}\,\text{Pa}$$

PRACTICE QUESTIONS

1 From the stress–strain graph on the previous page, calculate the Young modulus for:
  a the strong material
  b the ductile material.

2 From the stress–strain graph on the previous page, determine the ultimate tensile strength of:
  a the strong material
  b the ductile material
  c the plastic material.

3 Use the shape of the graph on the previous page to describe the behaviour of each material.

WORKED EXAMPLE

A steel wire is 6.0 m long and 1.6 mm in diameter. It is extended 1.0 mm by a force. The Young modulus of the steel is 2.0×10^{11} Pa.

Rearranging the equations: force $= \sigma A$ and $\sigma =$ Young modulus $\times \varepsilon$

so force $=$ Young modulus $\times \varepsilon A$

Strain, $\varepsilon = \dfrac{(1.0 \times 10^{-3}\,m)}{(6.0\,m)} = (1.67 \times 10^{-4})$

Area, $A = \pi r^2 = \pi(0.8 \times 10^{-3}\,m)^2 = 2.01 \times 10^{-6}\,m^2$

Force $=$ Young modulus $\times \varepsilon A$

Force $= (2.0 \times 10^{11}\,Pa)(1.67 \times 10^{-4})(2.01 \times 10^{-6}\,m^2) = 67\,N$

PRACTICE QUESTIONS

4 A piano wire made of steel (Young modulus $= 2.0 \times 10^{11}$ Pa) has a length of 1.2 m and a diameter of 1.8 mm. It stretches 2.5 mm when in tension. Calculate the tension in the wire.

5 Find the extension of a copper wire of length 2.0 m and diameter 3.2 mm when a force of 30 N is applied.
(Young modulus for copper $= 110$ GPa.)

STRETCH YOURSELF!

Calculate the minimum cross-sectional area a steel wire can have to safely suspend the weight of a 70 kg person without the elastic limit being exceeded. (The elastic limit of the steel is 5.0×10^8 Pa.)

5 ELECTRICITY

5.1 Resistance and resistivity

Resistance

The resistance, R, of a wire or component is defined by the equation:

$$\text{Resistance} = \frac{\text{potential difference (pd)}}{\text{electric current}} \qquad R = \frac{V}{I} \qquad \text{The unit is the ohm, } \Omega.$$

When $R = 1\,\Omega$, a pd of 1V causes a current of 1A.

This is true for any component because it is the way resistance is defined. Some components and metal wires obey Ohm's law:

The current through the component is directly proportional to the potential difference across it, as long as the temperature remains constant.

$V \propto I$ or $V = IR$ where R is the constant.

If a graph of current against potential difference is a straight line through the origin, the resistance is constant and the component obeys Ohm's law. If the graph is a curve, the component does not obey Ohm's law (sometimes described as non-ohmic). The resistance is still potential difference divided by current at any point, but this value is not a constant.

WORKED EXAMPLE

Resistance from a graph

On this graph current, I, is plotted against pd, V, because pd is the independent variable. It is sometimes called an IV characteristic.

The value of the resistance $R = \dfrac{1}{\text{gradient}}$

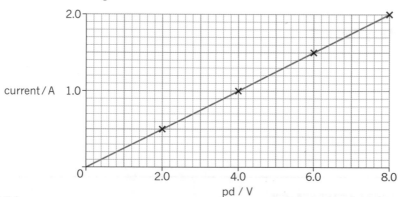

$\text{Gradient} = \dfrac{2.0\,\text{A}}{8.0\,\text{V}} = 0.25\,\Omega^{-1}$

$\text{Resistance} = \dfrac{1}{\text{gradient}} = 4.0\,\Omega$

(V against I graphs are also sometimes plotted. In this case, resistance = gradient.)

WORKED EXAMPLE

Resistors can be joined in series:

$R_1 = 3300\,\Omega \qquad R_2 = 100\,\Omega \qquad R_3 = 4700\,\Omega$

The total resistance $R_T = R_1 + R_2 + R_3$

$$R_T = 3300\,\Omega + 100\,\Omega + 4700\,\Omega$$
$$= 8100\,\Omega$$

WORKED EXAMPLE

Resistors can be joined in parallel, as shown by the diagram:

The total resistance is given by: $\frac{1}{R_T} = \frac{1}{R_1} + \frac{1}{R_2} + \frac{1}{R_3}$

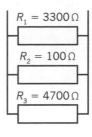

$R_1 = 3300\,\Omega$

$R_2 = 100\,\Omega$

$R_3 = 4700\,\Omega$

$$\frac{1}{R_T} = \frac{1}{3300\,\Omega} + \frac{1}{100\,\Omega} + \frac{1}{4700\,\Omega} = \frac{47 + 1551 + 33}{(3300 \times 47)\,\Omega}$$

$$\frac{1}{R_T} = \frac{47 + 1551 + 33}{155\,100\,\Omega} = \frac{1631}{155\,100\,\Omega}$$

$$R_T = \frac{155\,100\,\Omega}{1631} = 95\,\Omega \text{ (2 s.f.)}$$

(Notice that in parallel circuits the total resistance is always less than the smallest resistance as there are more paths for the current to pass through.)

PRACTICE QUESTIONS

1 Calculate the missing values:

  a $I = 200\,\text{mA}$ $R = ?$ $V = 12\,\text{V}$ **b** $I = ?$ $R = 4.7$ $\text{k}\Omega$ $V = 230\,\text{V}$

  c $I = 1.8\,\text{A}$ $R = 5.0\,\Omega$ $V = ?$

2 Calculate the total resistance of these resistors connected:

  a in series   **b** in parallel

  i $200\,\Omega$, $300\,\Omega$, $600\,\Omega$ **ii** $1\,\text{M}\Omega$, $100\,\text{k}\Omega$, $10\text{k}\Omega$ **iii** $0.5\,\Omega$, $1.0\,\Omega$, $2.0\,\Omega$

Resistivity

The resistivity, ρ, of a material is the resistance of a sample 1 m long with a cross–sectional area of $1\,\text{m}^2$.

$$\rho = \frac{RA}{L}$$

Where R is the resistance, A is the cross-sectional area, and L is the length.

Units are $\Omega\,\text{m}$ (not Ω per m).

WORKED EXAMPLE

The resistance of a sample of copper 2.0 m long and with a $0.50\,\text{mm}^2$ cross-sectional area (resistivity of copper $= 1.7 \times 10^{-8}\,\Omega\,\text{m}$) is:

$$R = \frac{\rho L}{A} = \frac{(1.7 \times 10^{-8}\,\Omega\,\text{m})}{(0.50 \times 10^{-6}\,\text{m}^2)} \times (2.0\,\text{m}) = 0.068\,\Omega$$

PRACTICE QUESTIONS

3 Calculate the resistivity of a metal wire with resistance $2.5\,\Omega$, length 1.8 m, and diameter 0.40 mm.

4 Calculate the resistance of a copper rod of length 80 cm and diameter 0.5 mm.
(Resistivity of copper $= 1.7 \times 10^{-8}\,\Omega\,\text{m}$.)

5 **a** Determine the length of nichrome wire of cross-sectional area $2.0 \times 10^{-7}\,\text{m}^2$ required to make a $10.0\,\Omega$ resistor.

  b How does this compare with tungsten?
(Resistivity of nichrome $= 1.1 \times 10^{-6}\,\Omega\,\text{m}$, resistivity of tungsten $= 5.6 \times 10^{-8}\,\Omega\,\text{m}$.)

5.2 Electric charge and current

Electric charge

Electric charge is positive or negative. A flow of charge is an electric current. When there is a current, I, for a small time interval, $\triangle t$, the charge that flows is: $\triangle Q = I\triangle t$.

Charge, Q, is measured in coulombs, C. $\quad 1\,C = 1\,As$

Electrons have a negative charge. The flow of electrons is in the opposite direction to the conventional current.

WORKED EXAMPLE

The charge on an electron is $e = -1.60 \times 10^{-19}\,C$.

The number of electrons, N, that are needed to make a charge of $-1.00\,C$ is:

$$N = -\frac{1.00\,C}{(-1.60 \times 10^{-19}\,C)} = 6.25 \times 10^{18}$$

If 1.0×10^6 electrons pass a point in a circuit in $0.5\,s$, the average current is:

$$I = \frac{1.0 \times 10^6 \times (-1.60 \times 10^{-19}\,C)}{(0.5\,s)} = -3.2 \times 10^{-13}\,A$$

(The negative sign shows current direction is opposite to electron flow.)

PRACTICE QUESTIONS

1   Calculate the number of electrons passing each point in a circuit in 1 minute when the current is $20\,mA$.

2   A charge of $6.0\,mC$ passes each point in a circuit in $2.0\,s$. Determine the current in the wire.

WORKED EXAMPLE

Mean drift velocity

A metal has free electrons that move when a potential difference (pd) is applied. This means that charge flows — there is a current. The free electrons are the charge carriers. They are moving with a mean drift velocity, v. The number of free electrons per m^3 is the number density of free electrons, n. It is different for different metals.

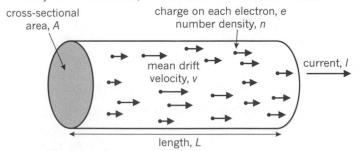

cross-sectional area, A

charge on each electron, e
number density, n

mean drift velocity, v

current, I

length, L

Electrons in a wire

In the wire the number of free electrons = number per unit volume × volume = nAL

The total charge carried by the electrons is $nALe$.

Current, $I = \dfrac{\text{total charge}}{\text{time}} = \dfrac{nALe}{t}$ and mean drift velocity, $v = \dfrac{L}{t}$

So $I = nAve$

For copper, $n = 8 \times 10^{28}\,\text{m}^{-3}$. If a copper wire has a cross-sectional area of $0.50\,\text{mm}^2$ and carries a current of $1.0\,\text{A}$ the drift velocity is:

$$v = \frac{1.0\,\text{A}}{(8 \times 10^{28}\,\text{m}^{-3})(0.50 \times 10^{-6}\,\text{m}^2)(1.60 \times 10^{-19}\,\text{C})} = 1.6 \times 10^{-4}\,\text{m s}^{-1}$$

Note that the magnitude of the charge, e, is usually used in calculations, but remember that for negative charge carriers (free electrons) v is in the opposite direction to I.

PRACTICE QUESTIONS

3 A copper wire of diameter $0.60\,\text{mm}$ carries a current of $2.0\,\text{A}$. The copper contains 8.0×10^{28} free electrons per m^3. Calculate the drift velocity of the electrons.

4 A metal wire contains 2.5×10^{28} free electrons per m^3 and has a diameter of $1.0\,\text{mm}$. Calculate the current in the wire if the electrons travel with a mean drift velocity of $0.50\,\text{mm s}^{-1}$.

5 An aluminium wire has a cross-sectional area of $4.0 \times 10^{-6}\,\text{m}^2$ and carries a current of $5.0\,\text{A}$. If there are 6.0×10^{28} free electrons per m^3, calculate the mean drift velocity of the electrons.

Current in circuits

In a complete circuit charge is never gained or lost; charge is conserved. At any junction the total current flowing into the junction = the total current flowing out of the junction.

WORKED EXAMPLE

In this example the current flowing into junction X is $6\,\text{A}$ and the current flowing out is $4\,\text{A} + I_A$

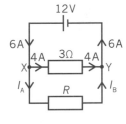

$6\,\text{A} = 4\,\text{A} + I_A$ so $I_A = 2\,\text{A}$

Similarly, at Y, $6\,\text{A} = 4\,\text{A} + I_B$ so $I_B = 2\,\text{A}$

Now the current in the unknown resistor is known, $R = \dfrac{V}{I}$ gives $R = \dfrac{12\,\text{V}}{2\,\text{A}} = 6\,\Omega$.

PRACTICE QUESTIONS

6 Calculate the missing values of current in this circuit:

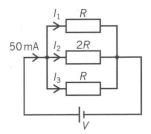

7 Calculate the missing values of current and resistance in this circuit:

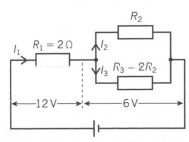

5.3 emf and potential difference

Energy, power, and pd

When electric charge flows through a component, energy is transferred from the charge to the component. The difference in energy per unit charge between when the charge enters and leaves the component is called the potential difference (pd) and is measured in volts (V), where $1V = 1JC^{-1}$ (1 joule of energy per coulomb of charge).

WORKED EXAMPLE

The energy transferred, and so the work done, W, when a charge, Q, passes through a component with a pd across it, V, is given by: $W = QV$.

When a charge of 25 C flows through a resistor and the pd across the resistor is 12 V, the energy transferred is equal to the work done, $W = (25\,C)(12\,V) = 300\,J$.

The power transferred is the energy transferred per second and the charge transferred per second is the current, so we also have the equation: $P = IV$.

If in the example above the energy is transferred in 5.0 s, the current $I = \dfrac{\Delta Q}{\Delta t} = \dfrac{(25\,C)}{(5.0\,s)} = 5\,A$ and the energy transferred $P = (5\,A)(12\,V) = 60\,J$.

For a resistor, R, using $V = IR$ in the equation $P = IV$ we get more equations for power:

$P = I(IR)$ gives $P = I^2R$

and $P = \dfrac{V}{R}V$ gives $P = \dfrac{V^2}{R}$

PRACTICE QUESTIONS

1 A 0.60 kW toaster is connected to a 230 V power source. Determine:
   a the current
   b the resistance.

2 The current supplied from the 230 V mains to a TV is 0.40 A. Calculate how much energy is transferred in an hour.

3 A CD player has current of 330 mA at 9.0 V. Determine the power it dissipates.

4 A car headlight is rated at 12 V and 75 W. It is connected to a battery with 88 ampere-hours of charge. Calculate how long it will take to completely discharge the battery.

---

emf and internal resistance

The electromotive force (emf), ε, is the energy transferred to the electric charge by the energy source, that is, the cell, battery, or power supply. It is a measure of the energy, E, given to the charge, Q, and, like pd, it is measured in volts.

$$\varepsilon = \frac{E}{Q}$$

All the energy is transferred from the charge to the wires and components of a circuit so the emf of the battery is numerically equal to the total pd across the circuit.

Up to now, you have assumed that all the energy is transferred from the battery to the circuit. In fact the output pd of the battery between its terminals is less than the emf because of the internal resistance, r, of the battery.

This can be treated as an extra resistance in the circuit, so instead of $\varepsilon = IR$ you have $\varepsilon = IR + Ir$ or emf, $\varepsilon = I(R + r)$

Using $V = IR$ gives $\varepsilon = V + Ir$ and rearranging this gives $V = \varepsilon - Ir$

A graph of the terminal pd, V (between points A and B), against the current, I, will give a straight line. It is of the form $y = mx + c$ where m is a negative value so the graph has a negative gradient, which is $-r$.

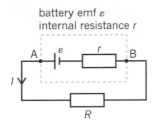

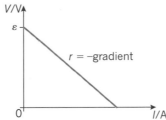

When I is zero $V = \varepsilon$, the pd across the terminals of the battery when it is an open circuit.

WORKED EXAMPLE

A battery has emf $\varepsilon = 12.5$ V and is connected to a resistor of 20 Ω. The terminal pd, V, is 12 V. You draw a circuit diagram, showing the internal resistance, r, in series with the battery.

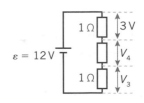

$I = \dfrac{V}{R} = \dfrac{(12\,V)}{(20\,\Omega)} = 0.60\,A$

$\varepsilon = V + Ir$ so $(12.5\,V) = (12\,V) + (0.60\,A)r$

$\dfrac{(12.5\,V) - (12\,V)}{(0.60\,A)} = r = 0.83\,\Omega$ The internal resistance is 0.83 Ω.

Note: $\varepsilon - V = (12.5\,V) - (12\,V) = 0.5$ V is sometimes called the 'lost volts'.

PRACTICE QUESTIONS

5 Calculate the missing values of pd in each of these circuits (internal resistance is negligible).

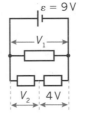

6 A battery of emf 9.0 V and internal resistance 0.50 Ω is connected to a 4.0 Ω resistor. Calculate the current through the resistor and the pd across it.

7 A battery of emf 12.0 V and internal resistance 0.50 Ω is connected in series to a 10.0 Ω resistor and an unknown resistor R. The pd across R is 6.0 V. Calculate:
 a the current through the resistor
 b the pd across it
 c the 'lost volts'.

8 Use these values for a battery to plot a graph and find the emf and the internal resistance:
 a $I = 2.0\,A$, $V = 11.0\,V$
 b $I = 4.0\,A$, $V = 10.0\,V$
 c $I = 8.0\,A$, $V = 8.0\,V$)
 d $I = 12.0\,A$, $V = 6.0\,V$
 e $I = 18.0\,A$, $V = 3.0\,V$.

5.4 The potential divider and other circuits

The potential divider

If you have a 9 V battery and you need a 6 V supply you might use a potential divider to give the 6 V you need. The output pd, V_{out}, is a fraction of the input pd, V_{in}. The fraction depends on the choice of the resistors R_1 and R_2. Often one of these is a variable resistor so it can be adjusted.

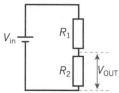

In the input loop the total resistance $= R_1 + R_2$.

So $V_{in} = I(R_1 + R_2)$

The pd V_{out} across R_2 is $V_{out} = IR_2$

Dividing these two equations gives: $\dfrac{V_{out}}{V_{in}} = \dfrac{R_2}{(R_1 + R_2)}$ or $V_{out} = \dfrac{R_2}{(R_1 + R_2)} V_{in}$

Also $\dfrac{V_{R1}}{V_{R2}} = \dfrac{R_1}{R_2}$

Note: if a resistance is connected across V_{out}, it must be about 10 times the value of R_2 or larger, or V_{out} is reduced.

 WORKED EXAMPLE

Calculate the value of V_{out} acheived with the potential divider in the circuit shown.

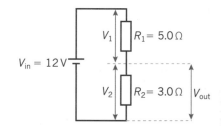

$V_{out} = \dfrac{(3.0\,\Omega)}{(5.0\,\Omega) + (3.0\,\Omega)} \times (12\text{ V}) = 4.5\text{ V}$

 WORKED EXAMPLE

The resistance of a light-dependent resistor decreases with increasing light intensity. This circuit can be used to switch on a light when the light intensity drops below a certain value.

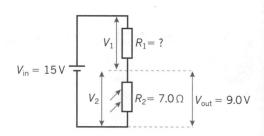

If the light must switch on at twilight, when the resistance of the LDR has a value of 7.0 kΩ and 9.0 V is required to switch on the light:

$V_{R2} = V_{out} = 9.0\text{ V} \quad V_1 = V_{in} - 9.0\text{ V} = 6.0\text{ V}$

$\dfrac{V_{R1}}{V_{R2}} = \dfrac{R_1}{R_2}$

$\dfrac{(6.0\text{ V})}{(9.0\text{ V})} = \dfrac{R_1}{(7.0\text{ k}\Omega)}$

$R_1 = (7.0\text{ k}\Omega) \times 0.667 = 4.7\text{ k}\Omega$

PRACTICE QUESTION

1 The input pd to a potential divider is 9.0 V.
   a Calculate the output pd across a 1.0 kΩ resistor if the second resistor is 5 kΩ.
   b Calculate the output pd across a 330 Ω resistor if the second resistor is 990 Ω.
   c If the output pd across a 680 Ω resistor is 7.0 V, calculate the other resistor.

STRETCH YOURSELF!

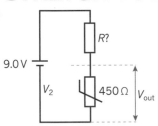

The resistance of a thermistor decreases with temperature. At 20 °C the resistance is 450 Ω. Using a supply of 9.0 V calculate the value of the series resistor required to switch on a heater at 20 °C using a pd of 5.0 V.

Solving circuit problems

If there is no circuit diagram, draw one and label all the unknown values you need to find. Then use these rules: (There is often more than one way to calculate the answer.)

1 In a series circuit, the emf, ε, is equal to the sum of the pds in the circuit (including the pd across the internal resistance if this is not negligible).

2 The pd across components in parallel is the same.

3 The total current entering a junction is equal to the total current leaving it.

4 The total resistance of a combination of resistors can be found using the formulae for resistors in series and in parallel (see Topic 5.1).

5 For resistors in series, the ratio of the pds across the resistors equals the ratio of the resistance values (as shown for the potential divider on the previous page).

PRACTICE QUESTIONS

2 Calculate the resistance, R, in this circuit:

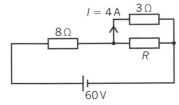

3 A parallel combination of a 1.0 Ω, a 3.0 Ω, and a 6.0 Ω resistor is connected in series with a parallel combination of three 2.0 Ω resistors. They are connected to a 1.5 V battery.

   Calculate:
   a the total current in the circuit
   b the current in each of the 2.0 Ω resistors
   c the current in the 3.0 Ω resistor.

6 QUANTUM PHYSICS

6.1 The photoelectric effect

Energy of a photon

In some situations electromagnetic radiation behaves as discrete 'packets' of energy called photons. The energy of a photon, E, is:

$$E = hf$$

Where f is the frequency of the electromagnetic radiation in Hz and $h = 6.63 \times 10^{-34}\,\text{J s}$ (the Planck constant).

As the speed of the radiation is $c = f\lambda$ (see Topic 2.1)

$$E = \frac{hc}{\lambda}$$

WORKED EXAMPLE

The energy of a photon of red light with frequency $4.3 \times 10^{14}\,\text{Hz}$ is:

$$E = (6.63 \times 10^{-34}\,\text{J s}) \times (4.3 \times 10^{14}\,\text{Hz}) = 2.9 \times 10^{-19}\,\text{J}$$

The energy of a photon of violet light with wavelength 350 nm is:

$$E = \frac{(6.63 \times 10^{-34}\,\text{Js}) \times (3.00 \times 10^{8}\,\text{ms}^{-1})}{(350 \times 10^{-9}\,\text{m})} = 5.7 \times 10^{-19}\,\text{J}$$

This is such a small number of joules that the electronvolt (eV) is often used as a unit of energy. This is the energy transferred when an electron moves through a pd of 1 V. Using the equation $W = QV$, Q = the charge on the electron $e = 1.60 \times 10^{-19}\,\text{C}$:

$$1\,\text{eV} = (1.60 \times 10^{-19}\,\text{C}) \times (1\,\text{V}) = 1.60 \times 10^{-19}\,\text{J}$$

The energy of the photon of red light above $= \dfrac{2.9 \times 10^{-19}\,\text{J}}{1.60 \times 10^{-19}\,\text{J per eV}} = 1.8\,\text{eV}$

PRACTICE QUESTIONS

1   Calculate the energy of the violet light photon in the worked example above, in eV.

2   Calculate the energy of a photon of yellow light of wavelength 590 nm in:

    a   J   **b**   eV

The photoelectric effect

When electromagnetic radiation falls on the surface of some materials, electrons are emitted. The electrons are called photoelectrons.

No electrons are emitted below a threshold frequency, f_0, which depends on the material. Below f_0 increasing the intensity of the radiation makes no difference, but above f_0 it increases the number of electrons emitted per second.

Above f_0 the maximum kinetic energy of the electrons emitted increases with frequency but is not affected by intensity. These observations cannot be explained by wave theory and are good evidence for each photon giving one electron a quantum of energy, $E = hf$. The minimum energy needed to free an electron from the surface is the work function $\phi = hf_0$.

$$hf = \phi + \tfrac{1}{2}m_e v_{max}^2$$

(The energy of the incident photon = minimum energy needed to free the electron from the surface + the kinetic energy of the emitted electron.)

If a circuit is set up so that radiation of frequency f falls on a photocell and a stopping potential, V_s is adjusted to the pd needed to just reduce the current to zero by stopping all of the electrons emitted from the cell, then:

$$eV_s = \tfrac{1}{2}m_e v_{max}^2$$

So $hf = \phi + eV_s$ and using $\phi = hf_0$ gives $hf = hf_0 + eV_s$, which can be arranged as the equation of a straight line:

$$V_s = \frac{hf}{e} - \frac{hf_0}{e}$$

Plotting a graph of V_s against f will give a straight line with gradient $\dfrac{h}{e}$ and when $V_s = 0$, $f = f_0$.

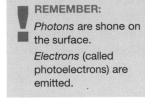

> **REMEMBER:**
> *Photons* are shone on the surface.
> *Electrons* (called photoelectrons) are emitted.

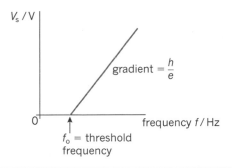

 WORKED EXAMPLE

This graph is the result of an experiment to determine the Planck constant. The stopping potential V_S in volts is plotted against the frequency of the electromagnetic radiation in Hz:

$$\text{Gradient} = \frac{(3.0\,\text{V} - 0\,\text{V})}{(12.0 - 4.4) \times 10^{14}\,\text{Hz}}$$

$$= \frac{3.0\,\text{V}}{7.6 \times 10^{14}\,\text{Hz}} = 3.95 \times 10^{-15}\ \text{V s}$$

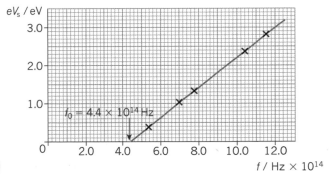

Gradient $= \dfrac{h}{e}$ so $h = (3.95 \times 10^{-15}\,\text{V s}) \times (1.60 \times 10^{-19}\,\text{C})$
which gives $h = 6.3 \times 10^{-34}\,\text{J s}$.

The Planck constant from the graph $= 6.3 \times 10^{-34}\,\text{J s}$ and the accepted value is $6.63 \times 10^{-34}\,\text{J s}$.

The percentage error in the experimental value is the error in the value, divided by the value, $\times 100\%$:

$$\frac{(6.63 - 6.3) \times 10^{-34}\,\text{J s}}{6.63 \times 10^{-34}\,\text{J s}} \times 100\% = 5.0\%\ (2\ \text{s.f.})$$

PRACTICE QUESTIONS

3 The work function for sodium is 2.46 eV. Calculate:

   a the lowest frequency that will cause emission of photoelectrons from the surface of sodium

   b the maximum kinetic energy of emitted photoelectrons with light of wavelength 450 nm.

4 When caesium metal is illuminated with light of wavelength 290 nm the maximum kinetic energy of the emitted electrons is 2.19 eV. Calculate:

   a the work function of caesium

   b the maximum kinetic energy of the emitted electrons with light of 350 nm.

6.2 Energy, waves, and particles

Energy levels

Electrons in atoms can only have certain discrete amounts of energy. These are called energy levels. To move from one level to another an electron must emit or absorb a photon with energy that exactly corresponds to the energy difference, ΔE in joules (J), between the two levels E_1 and E_2. The energy difference determines the frequency, f in hertz (Hz), of the photon absorbed or emitted,

$\Delta E = E_2 - E_1 = hf$ where the Planck constant $h = 6.63 \times 10^{-34}$ J s.

When the electron absorbs just enough energy to escape the atom, ionising it, the electron then has zero kinetic energy, so this level is defined as the zero energy level. If the electron is in one of the lower energy levels, bound to the atom, then its energy is less, so these energy levels are negative. (Compare with the Celsius temperature scale: if the temperature is just enough to melt ice we define that as $0\,°C$, if the temperature is less then we say it is negative.)

Each type of atom has a different set of energy levels because they depend on the charge, mass, and shape of the nucleus.

Energies are very small so are usually given using the electronvolt (eV) (see Topic 6.1)

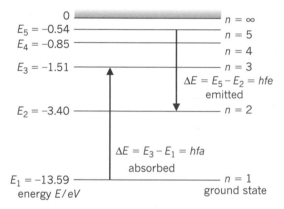

Some of the energy levels for the hydrogen atom

REMEMBER:
To change from eV to J **multiply** by the numerical value of the electron charge, $e = 1.6 \times 10^{-19}$.

To change from J to eV **divide**.

Which way round? The eV is for measuring very small energies – so the value of an amount of energy in J is much smaller than the value in eV.

WORKED EXAMPLE

A hydrogen atom is in its ground state. Calculate:
a the lowest frequency of light that can be absorbed
b all the photon energies in eV that can be emitted by an electron in state $n = 3$.

a Lowest f = smallest $\Delta E = E_2 - E_1 = (-3.40\ \text{eV}) - (-13.59\ \text{eV}) = 10.19\ \text{eV}$

$10.19\ \text{eV} = (10.19\ \text{eV})(1.6 \times 10^{-19}\ \text{J per eV}) = 1.63 \times 10^{-18}\ \text{J}$

$(1.63 \times 10^{-18}\ \text{J}) = (6.63 \times 10^{-34}\ \text{J s})f$

$f = 2.46 \times 10^{15}\ \text{Hz}$

b $\Delta E = E_3 - E_1 = (-1.51\ \text{eV}) - (-13.59\ \text{eV}) = 12.08\ \text{eV}$

$\Delta E = E_3 - E_2 = (-1.51\ \text{eV}) - (-3.40\ \text{eV}) = 1.89\ \text{eV}$

PRACTICE QUESTIONS

1 Use the diagram above to calculate the frequency required to ionise a hydrogen atom in the ground state.

2 a Sketch an energy level diagram for an atom with states $E = 0$, -2.7 eV, -5.5 eV, and -10.4 eV.

   b Show on the diagram all the transitions that result in a photon being emitted by the excited atom or ion.

   c Calculate all the photon frequencies.

   d Identify which of these can be absorbed by the atom in the ground state.

Wave particle duality

Light shows wave behaviour, such as diffraction and interference (see Topic 2.3), and particle behaviour, for example, in the photoelectric effect and emission and absorption in atoms (see Topic 6.1 and above). Similarly, small particles, such as electrons, which usually show particle behaviour, can also show wave behaviour. Electrons can be diffracted through gaps that are of the same order as the electron wavelength.

The de Broglie equation

Small particles can behave as waves with a wavelength called the de Broglie wavelength, λ, in metres (m).

$\lambda = \dfrac{h}{p}$ where the Planck constant $h = 6.63 \times 10^{-34}$ Js, and p is the momentum in N s.

When the particle is not moving at speeds close to the speed of light, $p = mv$ so

$\lambda = \dfrac{h}{mv}$

Often you know the kinetic energy $E_k = \frac{1}{2} mv^2$ so you can calculate the momentum:

$p = \sqrt{(m^2v^2)} = \sqrt{(m \times mv^2)} = \sqrt{(2m \times \frac{1}{2}mv^2)} = \sqrt{(2m \times E_k)}$

WORKED EXAMPLE

In an electron microscope the electrons are accelerated by a pd of 99 kV. Calculate the wavelength of the electrons.

When electrons are accelerated by a pd of 99 kV their energy is 99 keV.

99 keV = $(99 \times 10^3$ eV$)$ $(1.6 \times 10^{-19}$ J per eV$)$ OR   use $E = qV = (1.6 \times 10^{-19}$ C$)(99 \times 10^3$ V$)$

$E = 99 \times 10^3 \times 1.6 \times 10^{-19}$ J $= 1.58 \times 10^{-14}$ J

Using $p = \sqrt{(2m \times E_k)}$ you have $\lambda = \dfrac{h}{\sqrt{(2m \times E_k)}} = \dfrac{(6.63 \times 10^{-34}\,\text{J s})}{\sqrt{[2 \times (9.11 \times 10^{-31}\,\text{kg}) \times (1.58 \times 10^{-14}\,\text{J})]}} = 3.9 \times 10^{-12}$ m

Note that this is less than the wavelength of visible light, which is an advantage of electron microscopes; they can image smaller structures that would diffract light.

PRACTICE QUESTIONS

3 Calculate the wavelength of electrons with speed of 450 km s^{-1}.

4 Calculate the accelerating pd required to produce electrons with wavelengths of 100 picometres.

5 Neutrons are used for diffraction experiments. Calculate the momentum and energy of neutrons with a wavelength of the same order as atomic spacing of 2.6×10^{-10} m.

1 A stretched string 0.60 m long is fixed at both ends. It is plucked and set vibrating. Determine the wavelength of:

   a the 1st harmonic   **b** the 2nd harmonic   **c** the 3rd harmonic.

2 A student is using a diffraction grating to investigate the visible spectral lines from a mercury lamp.

   The green line has frequency of 5.49×10^{14} Hz and the first order green line is visible at an angle of 26° to the straight through position.

   a Calculate the grating spacing in lines per mm.

   b The first order yellow/orange line is visible at 27.5°. Calculate the wavelength of this line.

   c The blue line has a wavelength of 436 nm. Calculate the angle where the student should look for the second order blue line.

3 Car A and car B are both at the same place at time $t = 0$. The graph shows how their motion changes over the next 250 s.

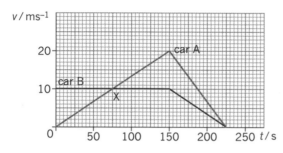

   a Calculate the final deceleration of car A.

   b Explain what happens at point X.

   c Use the graph to determine after how long car A overtakes car B. Explain your answer.

   d Calculate the total distance travelled by car A.

4

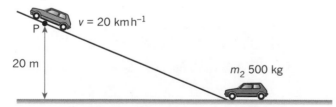

   A car of mass 700 kg is rolling down a straight road that is inclined at an angle, as shown in the diagram above. There is no driving force and the resistive forces are negligible. At point P it has a speed of 20 km h⁻¹ and is at a vertical height of 20 m above the bottom of the ramp. At the bottom of the ramp is a car of mass 500 kg. When the cars collide they move off together.

   a Calculate the velocity of the two cars immediately after the collision.

   The two cars now move along a horizontal road against a resistive force of 1500 N.

   b Determine how far they will move before they stop.

5 A hollow cube is made of steel. The outside measurement is 15 cm and the walls are 1 cm thick. Inside the cube is a sample of rock. The box and contents weighs 120 N. Calculate the mass of the rock.

The density of steel = 8050 kg m^{-3}

Gravitational field strength g = 9.81 N kg^{-1}

6 A steel wire is suspended vertically from a fixed hook and a 2.5 kg mass is attached to the end. The wire is 2.8 m long and has a diameter of 0.4 mm.

   a Calculate the final extension of the wire.

   b Calculate the energy stored in the wire.

   c State any assumptions made in your calculations.

The Young modulus for steel is 2.0 × 10^{11} Pa
Gravitational field strength, g = 9.81 N kg^{-1}

7 A circuit is connected as shown in the diagram. The battery has an internal resistance of 2.0 Ω. Resistors R_1 and R_2 are identical and have resistances of 15 Ω.

The lamp is operating at its normal operating conditions of 3.0 V and 600 mW.

Calculate:

   a the resistance of the lamp

   b the current through the lamp

   c the current through R_2

   d the potential difference across R_1

   e the emf of the supply.

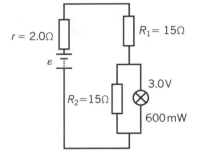

8 The work function of a zinc plate is 4.3 eV.

   a Calculate the threshold frequency of the zinc.

   b Calculate the maximum wavelength of the electromagnetic radiation that will cause emission of electrons from the zinc surface.

9 The diagram shows some energy levels of a mercury atom and some of the transitions that give spectral lines.

   a Calculate how much energy in eV is required to ionise the atom in its ground state.

   b Determine which of the lines X, Y, or Z shown in the diagram corresponds to the spectral line of wavelength 436 nm.

   c An atom in the ground state collides with an electron and gains 1.233 × 10^{-18} J of energy. It then emits a photon. Calculate the largest wavelength of radiation that it could emit, assuming there were no other energy levels.

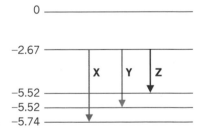

7.1 Circular motion

Radians

One radian is equal to the angle at the centre of a circle that subtends an arc equal to the radius.

The angle, θ, in radians is the length of the arc s divided by the radius of the circle r.　$\theta = \frac{s}{r}$　or　$s = r\theta$

For a whole circle the length of the arc is the circumference, $2\pi r$, so the angle θ in radians that is equal to 360° is $\theta = \frac{2\pi r}{r} = 2\pi$.

2π radians = 360° and π^c = 180° using c, which is the symbol for radians.

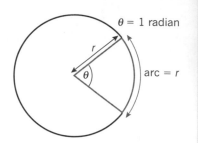

 WORKED EXAMPLE

To change from degrees to radians: $90° = \frac{(90° \times 2\pi)}{360°}$ radians $= \frac{\pi}{2}$ radians, this is called 'pi by two' radians = 1.57 radians.

To change from radians to degrees: $\frac{2\pi}{3}$ radians $= \frac{(2\pi \times 360°)}{(3 \times 2\pi)} = 120°$

 PRACTICE QUESTIONS

1 Give the following in radians. Give your answer as a multiple of π:

   a 60°   **b** 180°   **c** 540°

2 Give the following in degrees:

   a $\frac{\pi}{4}$ radians   **b** $\frac{3\pi}{2}$ radians   **c** $\frac{\pi}{6}$ radians

   d 1 radian   **e** 12.6 radians

Angular velocity

The diagram shows an object moving in a circle with radius r at velocity v.

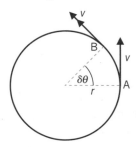

The period, T, is the time for one complete rotation in seconds.

Since the distance travelled in time T is $2\pi r$, the velocity, $v = \frac{2\pi r}{T}$.

The frequency, f, is the number of rotations per second in Hz. $f = \frac{1}{T}$.

The angular velocity, ω, is the angle turned through, $\delta\theta$, in a time δt.

$\omega = \frac{\delta\theta}{\delta t}$　angular velocity is measured in radians per second.

Since the distance travelled $s = r\theta$, dividing by the time taken gives the velocity $v = r\omega$.

The angle turned through in time T is 2π, the angular velocity $\omega = \frac{2\pi}{T}$, and $\omega = 2\pi f$.

WORKED EXAMPLE

A wheel turning at 12.0 radians in 2.0 s has an angular velocity of

$$\omega = \frac{(12.0 \text{ radians})}{(2.0\text{s})} = 6.0 \text{ radians s}^{-1}.$$

A wheel that makes five rotations per second ($f = 5\,\text{Hz}$) has $\omega = 2\pi(5\,\text{Hz}) = 10\pi$ radians s^{-1}.

 REMEMBER:
An object moving in a circle needs a centripetal force towards the centre of the circle. Think of an object on a string. The tension in the string provides the centripetal force. If the string breaks the object continues in a straight line — along a tangent. The centripetal force is the resultant of the forces on the object.

Centripetal force and acceleration

Centripetal acceleration $a = \dfrac{v^2}{r} = r\omega^2$ and using $F = ma$

Centripetal force $= \dfrac{mv^2}{r} = mr\omega^2$

WORKED EXAMPLE

A mass of 0.40 kg on the end of a string is moving in a horizontal circle. The circle makes an angle of 50° with the horizontal and the radius is 0.20 m.

Resolving (see Topic 3.5) and using $\dfrac{\sin\theta}{\cos\theta} = \tan\theta$

$\rightarrow F\cos 50° = \dfrac{mv^2}{r}$ **(1)** $\uparrow F\sin 50° = mg$ **(2)** where F = tension in the string

From **(2)** $F = \dfrac{(0.40\,\text{kg})(9.81\,\text{m s}^{-2})}{\sin 50°} = 5.1\,\text{N}$

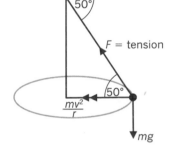

Dividing **(1)** and **(2)** $\dfrac{F\sin 50°}{F\cos 50°} = \tan 50° = \dfrac{mgr}{mv^2} = \dfrac{gr}{v^2}$

$v^2 = \dfrac{(9.81\,\text{m s}^{-2})(0.20\,\text{m})}{\tan 50°} = 1.64\,\text{m}^2\,\text{s}^{-2}$ giving $v = 1.3\,\text{m s}^{-1}$

PRACTICE QUESTIONS

3 An object moving in a circle has a period of 0.20 s. Calculate the angular velocity.

4 An object of mass 2.0 kg moves in a horizontal circle of radius 6.0 m with a constant speed of 12 m s^{-1}. Calculate:

   a the angular velocity

   b the centripetal force.

STRETCH YOURSELF!

An object of mass 6.0 kg is attached to the end of a string attached to a fixed point. It is moved in a vertical circle of radius 1.5 m at a constant speed of 5 m s^{-1}. Calculate the maximum and minimum tensions in the string.

7.2 SHM 1

Simple harmonic motion

When an object moves so that its acceleration a is directly proportional to its distance x from a fixed point and is always directed towards that point, this is called simple harmonic motion (SHM).

The equation describing the motion is always of the form: $a = -\omega^2 x$ where ω is a constant.

Using $\omega = 2\pi f$ the equation can also be written $a = -(2\pi f)^2 x$ where f is the frequency of the motion.

A useful example for showing the connection between the equations for circular motion and for SHM is the motion of the projection Q, onto the diameter, of a point P moving in a circle at constant speed.

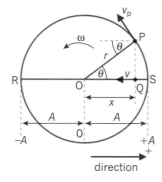

The point Q starts at S, the position where the displacement is a maximum, at time $t = 0$.

S is the maximum displacement from the fixed point O, when $x = A$, where A is the amplitude of the oscillation. (In this case $A = r$, the radius of the circle.)

The point Q moves:

- from S towards O increasing speed and reaching a maximum speed as it passes through O
- on towards R, slowing to a stop at R. At R, $x = -A$
- back towards O increasing speed and reaching a maximum speed as it passes through O
- on towards S, slowing to a stop at S. The motion then repeats.

From the movement of P around the circle, at any time $x = r\cos\theta$.

θ in radians is the angle moved through in time t when the angular velocity is ω, so, $\theta = \omega t$. This gives the displacement of the point Q:

$x = A\cos\omega t$ or $x = A\cos(2\pi f)t$

The velocity v_P of P around the circle is $v_P = r\omega$. The velocity of Q is the horizontal component of v_P.

$v = v_P\sin\theta$ in triangle OPQ: $\sin\theta = \pm\dfrac{\sqrt{(r^2 - x^2)}}{r}$ so $v = \pm r\omega\dfrac{\sqrt{(r^2 - x^2)}}{r} = \pm\omega\sqrt{(r^2 - x^2)}$

The velocity of Q: $v = \pm\omega\sqrt{(A^2 - x^2)}$ or $v = \pm(2\pi f)\sqrt{(A^2 - x^2)}$

The negative values of velocity are when Q is between O and R and is moving back towards O. Some of these equations, or similar ones, are given on the formulae sheet you are given in the exams.

> **REMEMBER:**
> For SHM and circular motion the angle θ must be in radians, not degrees.
>
> If the angle is given in degrees you must change it to radians.
>
> Make sure your calculator is set to radians to work out trig functions such as sines or cosines.

WORKED EXAMPLE

An object moves with SHM. It has an amplitude A of 2.0 cm and a frequency of 24 Hz.

The period of oscillation is: $T = \dfrac{1}{f} = \dfrac{1}{(24\,Hz)} = 0.0417\,s = 0.42\,s$ (2 s.f.)

The acceleration at the middle of the oscillation (when $x = 0$) is 0 ($a = -(2\pi f)^2 x = 0$)

The velocity at this point is: $v = \pm(2\pi f)\sqrt{(A^2 - x^2)} = \pm 2\pi f A = \pm 2\pi(24\,Hz)(2.0 \times 10^{-2}\,m)$
$$= 3.0\,m\,s^{-1}$$

The acceleration at the end of an oscillation (when $x = A$) is $a = -[2\pi(24\,Hz)]^2(2.0 \times 10^{-2}\,m)$
$$= 455\,m\,s^{-2}$$

The velocity at this point is: $v = \pm(2\pi f)\sqrt{(A^2 - x^2)} = \pm(2\pi f)\sqrt{(A^2 - A^2)} = 0$

WORKED EXAMPLE

A steel strip is clamped at one end. It vibrates with SHM and the free end has an amplitude, A, of 8.0 mm and a frequency of 50.0 Hz.

The period of oscillation is: $T = \dfrac{1}{f} = \dfrac{1}{(50.0\,Hz)} = 0.020\,s$

The velocity when passing through the equilibrium position is:
$v = \pm(2\pi f)\sqrt{(A^2 - x^2)} = \pm 2\pi f A = \pm 2\pi(50.0\,Hz)(8.0 \times 10^{-3}\,m) = 2.5\,m\,s^{-1}$

The acceleration at the maximum displacement (when $x = A$) is:
$a = -[2\pi(50.0\,Hz)]^2(8.0 \times 10^{-3}\,m) = 790\,m\,s^{-2}$

The displacement 2.5 m s⁻¹ after being released from the maximum displacement is:
$x = A\cos(2\pi f)t = (8.0 \times 10^{-3}\,m)\cos[(2\pi)(50.0\,Hz)(2.5 \times 10^{-3}\,s)] = (8.0 \times 10^{-3}\,m)\cos(0.25\pi)$
$\cos(0.25\pi) = \cos(0.785\,radians) = 0.707$
$x = (8.0 \times 10^{-3}\,m)(0.707) = 5.7 \times 10^{-3}\,m$ or 5.7 mm

---

PRACTICE QUESTIONS

1 A particle oscillates with SHM of amplitude 48 mm and maximum speed of 0.24 m s⁻¹. Calculate:

  a the period of the oscillation

  b the maximum acceleration.

2 The tip of a tuning fork vibrating at 512 Hz has a maximum speed of 4.2 m s⁻¹. Determine the amplitude.

3 A particle moves with SHM and its displacement x at time t is given by: $x = (2.0\,mm)\cos(3\pi t)$. Calculate:

  a the frequency   **b** the period   **c** the amplitude

  d the maximum speed   **e** the maximum acceleration.

  At time $t = 0.06\,s$ calculate:

  f the displacement   **g** the velocity   **h** the acceleration.

STRETCH YOURSELF!

A piston moves with SHM with amplitude of 6.0 cm and rotates a wheel at 80 rotations per second. It has a mass of 0.55 kg. Calculate the maximum force on it during the motion.

7.3 SHM 2

Graphs of SHM

To plot graphs of the displacement, velocity, and acceleration against time for a particle moving with SHM we use the following equations.

The equation for the displacement of a particle moving with SHM with amplitude A and frequency f is:

$x = A \cos \omega t$ where $\omega = 2\pi f$

The equation for the velocity is:

$v = \pm A \omega \sin \omega t$

(This is the same as $v = \pm\omega\sqrt{(A^2 - x^2)}$; see the horizontal component of v_P on Topic 7.2.)

The equation for the acceleration is:

$a = -A\omega^2 \cos \omega t$ (This is found by substituting $x = A \cos \omega t$ in $a = -\omega^2 x$)

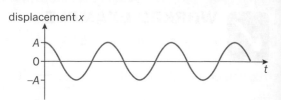

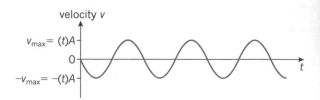

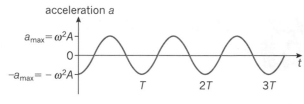

These equations and graphs are for SHM, which starts at $t = 0$ when $x = A$.

![checkmark] **WORKED EXAMPLE**

Plot a graph showing the displacement of a particle moving with SHM with $A = 5.0\,\text{cm}$ and $f = 10\,\text{Hz}$ starting at $t = 0$ when $x = 0$.

Notice that the graph is shifted back by $\dfrac{\pi}{2}$. $\cos \omega t = \sin(\omega t + \phi)$ where

$\phi = \dfrac{\pi}{2}$ ϕ is the phase angle (see Topic 2.2).

 PRACTICE QUESTIONS

1 Sketch graphs for the velocity and the acceleration of the particle in the worked example above.

2 Sketch displacement, velocity, and acceleration against time graphs for the SHM of a particle with frequency $f = 50\,\text{Hz}$ and $A = 2.0\,\text{mm}$ that starts from $2.0\,\text{mm}$ from the equilibrium position at time $t = 0$.

Energy transfer

A particle moving with SHM has kinetic energy $E_K = \frac{1}{2}mv^2 = \frac{1}{2}m(2\pi f)^2(A^2 - x^2)$.

As the particle passes through the equilibrium position $x = 0$ it has maximum velocity and no acceleration so there is no resultant force acting on it. At this point it has the maximum kinetic energy:

$E_K = \frac{1}{2}mv_{max}^2 = \frac{1}{2}m(2\pi fA)^2$

As the particle moves away from equilibrium and slows down, the kinetic energy is transferred to potential energy, E_P. At the furthest point of the oscillation when $x = \pm A$ the particle stops moving so $E_K = 0$ and E_P is a maximum. If the oscillation continues then energy loss is negligible and at any point the sum $E_K + E_P$ is a constant value for the total energy.

A graph of the energy against displacement for SHM:

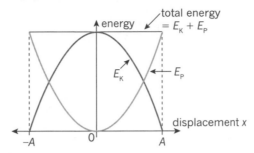

The equation for E_P depends on how the energy is stored.

For a pendulum, the bob rises as it swings, increasing the gravitational potential energy, $E_P = mgh$.

For a mass attached to a horizontal spring, the string stretches, increasing the elastic potential energy, $E_P = \frac{1}{2}kx^2$ (see Topic 4.1).

WORKED EXAMPLE

A pendulum bob of mass 150 g swings with an amplitude of 12 cm and a frequency of 1.0 Hz. Sketch a graph of the kinetic energy, the potential energy, and the total energy against time for one complete cycle.

The SHM has maximum E_K at the equilibrium position. This is equal to the total energy, and equal to the potential energy when $E_K = 0$ at $x = \pm A$

maximum $E_K = \frac{1}{2}m(2\pi fA)^2 = \frac{1}{2} \times (0.15\,\text{kg}) \times [2\pi(1.0\,\text{Hz})(0.12\,\text{m})]^2 = 0.043\,\text{J}$

$T = f^{-1} = \dfrac{1}{1.0\,\text{Hz}} = 1.0\,\text{s}$

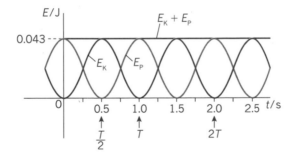

PRACTICE QUESTIONS

3 A pendulum bob of mass 250 g swings with an amplitude of 7.5 cm and a frequency of 5.5 Hz. Sketch a graph of the kinetic energy, the potential energy, and the total energy against time for one complete cycle.

4 A horizontal spring has a spring constant of $1.2 \times 10^2\,\text{N m}^{-1}$ and is attached to a mass. When it is stretched by 0.15 m and released it oscillates with SHM with a period of 0.2 s.

   a Determine the maximum potential energy.

   b Sketch graphs of the potential energy, kinetic energy, and total energy against displacement.

   c Sketch graphs of the potential energy, kinetic energy, and total energy against time for one cycle.

7.4 Examples of SHM

The pendulum

When a pendulum swings with a small amplitude and air resistance is negligible, it swings with SHM.

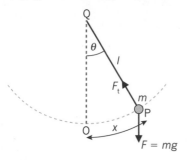
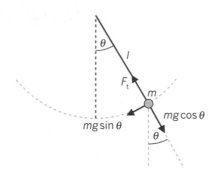

The mass m is displaced by x and the resultant force pulling it back towards the equilibrium position O is the component of its weight mg towards O:

$$F = mg \sin \theta$$

Notice that, if θ is in radians, $l \sin \theta$ is less than the arc length PO $= l\theta$, which is less than $l \tan \theta$, so:

$$\sin \theta < \theta < \tan \theta$$

As θ gets smaller and smaller, the three lengths get closer in value. This is called the 'small angle approximation'.

$$\sin \theta \approx \theta \approx \tan \theta \text{ and } \cos \theta \approx 1$$

So for very small angles $\sin \theta = \dfrac{x}{l}$

force towards O: $F = \dfrac{mgx}{l}$

Force in the positive x direction $= -\dfrac{mgx}{l}$

acceleration $= -\dfrac{gx}{l}$, which is the equation for SHM with $(2\pi f)^2 = \dfrac{g}{l}$

As the period $T = f^{-1}$ this can be rearranged to give $T = 2\pi \sqrt{\dfrac{l}{g}}$

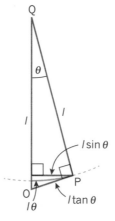

> **REMEMBER:** Because you have used the small angle approximation, pendulums must have a small amplitude to swing with SHM, less than about 10° (= 0.17 radians).

WORKED EXAMPLE

The period of a pendulum with length 0.75 m is:

$$T = 2\pi \sqrt{\frac{l}{g}} = 2\pi \sqrt{\frac{(0.75\,\text{m})}{(9.81\,\text{m s}^{-2})}}$$

$$T = 1.7\,\text{s}$$

WORKED EXAMPLE

The length of a pendulum with a period of 6.0 s would be:

$$(6.0\,\text{s}) = 2\pi \sqrt{\frac{l}{g}} =$$

$$l = \frac{(6.0\,\text{s})^2 \times (9.81\,\text{m s}^{-2})}{4\pi^2} = 8.9\,\text{m}$$

PRACTICE QUESTIONS

1 Show that the SHM equation derived above for the pendulum gives a period of $T = 2\pi\sqrt{\dfrac{l}{g}}$.

2 Determine the length of pendulum needed to give a period of 1.0 s.

A mass on a spring

When a mass is attached to a horizontal spring and is pulled and released, it will make small oscillations about the equilibrium position. If the friction is negligible it will oscillate with SHM.

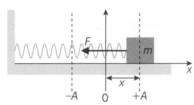

The mass is displaced by x and the resultant force pulling it back towards the equilibrium position O is $F = kx$ where k is the spring constant (see Topic 4.1).

In the positive x direction $F = -kx$

Acceleration, $a = -\dfrac{kx}{m}$

This is the equation for SHM with $(2\pi f)^2 = \dfrac{k}{m}$

$2\pi f = \dfrac{k}{m}$

$f = \dfrac{1}{2\pi}\sqrt{\dfrac{k}{m}}$

$T = 2\pi\sqrt{\dfrac{m}{k}}$

WORKED EXAMPLE

A mass of 0.15 kg oscillates on the end of a horizontal spring of stiffness 35 N m^{-1}. Find the period of the motion.

$m = 0.15\,\text{kg} \quad k = 35\,\text{N m}^{-1} \quad T = 2\pi\sqrt{\dfrac{m}{k}}$

$T = 2\pi\sqrt{\dfrac{(0.15\,\text{kg})}{(35\,\text{N m}^{-1})}} = 0.41\,\text{s}$

PRACTICE QUESTIONS

3 A mass of 0.64 kg oscillates on the end of a horizontal spring of stiffness 48 N m^{-1}. Determine the period of the motion.

4 A mass of 0.40 kg oscillates on the end of a horizontal spring with frequency of 2.0 Hz. Calculate:

    a the period of the motion

    b the spring constant of the spring.

8 THERMAL PHYSICS

8.1 Temperature scales and the gas laws

The kelvin scale

One kelvin (K) is the same size as one degree on the Celsius scale but its zero value is 'absolute zero', which is equivalent to $-273.15\,°C$. To convert a temperature from $\theta\,(°C)$ to $T\,(K)$:

$$T\,(K) = \theta\,(°C) + 273.15\,K$$

(Using 273 K is accurate enough in most cases.)

Unlike the Celsius scale, the kelvin scale is related to the average kinetic energy per molecule. All materials have minimum internal energy at absolute zero. This has the advantage that the temperature in kelvin is directly proportional to the average kinetic energy per molecule.

WORKED EXAMPLE

The temperature of boiling water is 100 °C. This is equivalent to:

$T\,(K) = 100\,(°C) + 273.15\,K = 373.15\,K.$

The temperature of melting ice is 273.15 K.

This is equivalent to: $\theta(°C) = T(K) - 273.15\,K = 273.15\,K - 273.15\,K = 0\,(°C).$

PRACTICE QUESTIONS

1   Liquid nitrogen boils at 77 K. State this temperature in °C.

2   Carbon dioxide sublimes (changes from solid to vapour) at $-78\,°C$. State this temperature in K.

> **REMEMBER:** The unit is kelvin – not degrees kelvin.

The gas laws

Boyle's law

For a fixed mass of gas at constant temperature, the pressure, p, is inversely proportional to the volume, V:

$p \propto \dfrac{1}{V}$ which means that $pV = $ constant

A graph of p against V is a curve but plotting p against $\dfrac{1}{V}$ gives a straight line through the origin.

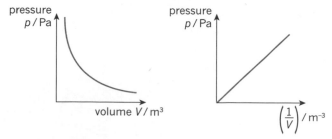
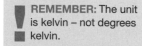

Charles's law

For a fixed mass of gas at constant pressure, the volume, V, is directly proportional to the temperature, T, in kelvin:

$V \propto T$ which means that $\dfrac{V}{T}$ is constant.

A graph of V against T is a straight line through the origin.

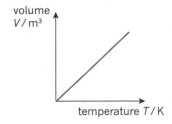

Charles's law predicts the gas will have zero volume at absolute zero. A real gas does not behave like this, so the law breaks down close to absolute zero.

The pressure law

For a fixed mass of gas at constant volume, the pressure, p, is directly proportional to the temperature, T, in kelvin:

$p \propto T$ which means that $\frac{p}{T}$ is constant.

A graph of p against T is a straight line through the origin.

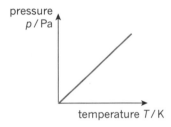

This law predicts that the pressure of an ideal gas is zero at absolute zero.

WORKED EXAMPLE

The pressure inside an airbed is 1.2×10^5 Pa and the volume is $0.15\,m^3$. A boy steps on the airbed and the volume reduces to $0.095\,m^3$. The temperature is unchanged. Calculate the new pressure.

$p_1V_1 = p_2V_2$

$(1.2 \times 10^5\,Pa)(0.15\,m^3) = p_2(0.095\,m^3)$

$p_2 = \dfrac{(1.2 \times 10^5)(0.15\,m^3)}{(0.095\,m^3)} = 1.9 \times 10^5\,Pa$

PRACTICE QUESTIONS

3 A bubble of air rises from the sea bed to the surface. Its pressure changes from 3.1×10^5 Pa to 1.0×10^5 Pa. The volume when it is on the sea bed is $4.2 \times 10^{-6}\,m^3$. Calculate the volume of the air at the surface.

4 The pressure inside a bicycle tyre is 6.0×10^5 Pa. It cools from 30 °C to 21 °C. If the volume stays constant, determine the new pressure.

5 A soft balloon has a volume of $0.011\,m^3$. It cools from 21 °C to 13 °C. If the pressure stays at atmospheric pressure, determine the new volume.

8.2 An ideal gas

The equation of state for an ideal gas

The gas laws on the previous page can be combined into one law for an ideal gas. An ideal gas is one that obeys Boyle's law. Real gases approximate to ideal gases providing they are well above the temperature when they liquefy and are at low densities. The equation of state for an ideal gas says, for a fixed mass of gas:

$$\frac{pV}{T} = \text{constant} \quad \text{or} \quad \frac{p_1V_1}{T_1} = \frac{p_2V_2}{T_2}$$

WORKED EXAMPLE

Calculate the volume of air at 100 °C and 201 kPa when it has a volume of 0.78 m³ at 0 °C and 101 kPa.

It is a good idea to write out all the variables you will be substituting to avoid mistakes. Add 273 to get T in K. (As the rest of the data is only to 2 s.f., the extra figures will make no difference to your answer):

$V_1 = 0.78\,\text{m}^3$ $T_1 = (0 + 273)\,\text{K}$ $p_1 = 101 \times 10^3\,\text{Pa}$

$V_2 = ?$ $T_2 = (100 + 273)\,\text{K}$ $p_2 = 201 \times 10^3\,\text{Pa}$

$$\frac{(101 \times 10^3\,\text{Pa})(0.78\,\text{m}^3)}{273\,\text{K}} = \frac{(201 \times 10^3\,\text{Pa})V_2}{373\,\text{K}}$$

$$V_2 = \frac{259 \times 373\,\text{m}^3}{201 \times 10^3} = 0.48\,\text{m}^3$$

PRACTICE QUESTIONS

1 A hydrogen gas balloon has a volume of 7200 m³ at 22 °C and the pressure inside is 1.0×10^5 Pa. It was filled from gas cylinders with volume of 25 m³ at 12 °C. Calculate the pressure in the gas cylinders.

2 The volume of a gas is halved whilst its pressure is tripled. Calculate the ratio of the final to the original temperature of the gas.

> **REMEMBER:** You must change the temperatures to kelvin.

STRETCH YOURSELF!

An ideal gas is in a closed tank at temperature of 15 °C and pressure of 1.01×10^6 Pa. The valve is opened and one half of the gas is removed. The temperature is increased to 55 °C. Calculate the new pressure in the tank.

The universal gas constant

The constant in the equation of state above depends on the mass of gas and also on the type of gas. However, if the amount of gas is 1 mole, then the constant is the same for all gases.

1 mole of a substance contains 6.02×10^{23} particles − which may be atoms, molecules, or other particles being considered.

The number $N_A = 6.02 \times 10^{23}\,\text{mol}^{-1}$ is the Avogadro constant. (This is the number of molecules in 0.012 kg of pure carbon-12, which is 1 mole of pure carbon-12.)

For 1 mole of any gas: $pV = RT$ where R is the molar gas constant.
$R = 8.31 \, \text{J} \, \text{K}^{-1} \, \text{mol}^{-1}$

For n moles of any gas: $pV = nRT$

The number of moles $n = \dfrac{m}{M}$ where m is the mass of gas and M is the molar mass (mass of 1 mole) in $\text{kg} \, \text{mol}^{-1}$.

REMEMBER:
When you are calculating the mass of a mole of particles, using relative atomic masses, you must take into account if the particles are atoms or molecules.

For example, the relative atomic mass of oxygen is 16:
- a mole of oxygen atoms has a mass of 0.016 kg
- a mole of oxygen molecules (O_2) has a mass of 0.032 kg.

WORKED EXAMPLE

A gas cylinder contains 3.5 kg of nitrogen gas at a pressure of 3.2×10^5 Pa. An identical cylinder contains 6.4 kg of oxygen at the same temperature. Calculate the pressure of the oxygen.

(Relative molecular mass: $O_2 = 32$ and $N_2 = 28$)

$V_1 = V_2$ $\qquad T_1 = T_2$ $\qquad p_1 = 3.2 \times 10^5$ Pa $\qquad p_2 - ?$

$m_1 = 3.5$ kg $\qquad M_1 = 0.028$ kg mol^{-1} $\qquad m_2 = 6.4$ kg $\qquad M_2 = 0.032$ kg mol^{-1}

$pV = nRT = \dfrac{mRT}{M}$ and V, R, and T are the same for both gases.

Rearranging the equation gives: $\dfrac{pM}{m} = \dfrac{RT}{V}$ so, $p_1 \dfrac{M_1}{m_1} = \dfrac{p_2 M_2}{m_2}$

$p_2 = p_1 \dfrac{M_1 m_2}{m_1 M_2}$

$p_2 = \dfrac{(3.2 \times 10^5 \, \text{Pa})(0.028 \, \text{kg mol}^{-1})(6.4 \, \text{kg})}{(3.5 \, \text{kg})(0.032 \, \text{kg mol}^{-1})} = 5.1 \times 10^5$ Pa

PRACTICE QUESTIONS

3 A sealed container with volume of 8.0×10^{-3} m^3 contains a gas with temperature of 25 °C and pressure of 9.0×10^5 Pa.

   a Calculate the number of moles of gas.

   b Calculate how many molecules of gas there are.

4 Show that the volume of 1 mole of gas at 0 °C and 1.013×10^5 Pa (standard atmospheric pressure and temperature) has a volume of about 22 litres.

5 An empty room has a volume of 27 m^3 and temperature of 22 °C. The air pressure is 1.01×10^5 Pa. Calculate the mass of air in the room.
(Assume relative atomic mass of air $= 0.029 \, \text{kg mol}^{-1}$.)

STRETCH YOURSELF!

Calculate the density of hydrogen gas at 25 °C and 1.01×10^5 Pa.

(The molar mass of hydrogen molecules is 0.002 kg mol^{-1}.)

8.3 Kinetic theory

The kinetic theory equation

In the kinetic theory, the pressure of a gas is treated as the sum of the forces exerted on the sides of a container by moving molecules (or atoms for monatomic gases like helium). The kinetic theory equation is:

$$pV = \tfrac{1}{3}Nm<c^2>$$

p is the pressure and V is the volume of the gas, N is the number of molecules, and m is the mass of a molecule (do not confuse this with the mass of the gas). The factor of $\tfrac{1}{3}$ comes from the fact that the molecules are moving in three dimensions.

$<c^2>$ is the mean square speed of the molecules $= \dfrac{(c_1^2 + c_2^2 + c_3^2 + \ldots + c_N^2)}{N}$

$c_{rms} = \sqrt{<c^2>}$ is the root mean square (rms) speed of the molecules

$$c_{rms} = \sqrt{\dfrac{(c_1^2 + c_2^2 + c_3^2 + \ldots + c_N^2)}{N}}$$

Note that $c_{rms}^2 = <c^2>$ so the kinetic theory equation is sometimes written:

$$pV = \tfrac{1}{3}Nmc_{rms}^2$$

> **REMEMBER:** The rms speed is not the same as the mean speed.
> For example, mean speed:
> $= \tfrac{1}{5}(1.0\,ms^{-1} + 3.0\,ms^{-1} + 3.0\,ms^{-1} + 5.0\,ms^{-1} + 7.0\,ms^{-1})$
> $= 3.8\,ms^{-1}$
>
> But rms speed:
> $= \sqrt{\tfrac{1}{5}(1.0\,m^2s^{-2} + 9.0\,m^2s^{-2} + 9.0\,m^2s^{-2} + 25\,m^2s^{-2} + 49\,m^2s^{-2})} = 4.3\,ms^{-1}$

WORKED EXAMPLE

At 4 °C and atmospheric pressure of $1.01 \times 10^5\,Pa$ the density of air is about $0.99\,kg\,m^{-3}$. Calculate the root mean square speed of the air molecules.

$pV = \tfrac{1}{3}Nmc_{rms}^2$

Using density $\rho = \dfrac{Nm}{V}$ gives: $p = \tfrac{1}{3}\rho c_{rms}^2$

$c_{rms}^2 = \dfrac{3p}{\rho} = \dfrac{3(1.01 \times 10^5\,Pa)}{(0.99\,kg\,m^{-3})} = 3.06 \times 10^5\,m^2\,s^{-2}$

$c_{rms} = 550\,m\,s^{-1}$

PRACTICE QUESTION

1   This table shows the number of particles N with speed v.

N	6	9	10	12	13	10	7	5
v/ms^{-1}	10.0	20.0	30.0	40.0	50.0	60.0	70.0	80.0

Calculate: **a** the mean speed   **b** the rms speed.

2   Calculate the rms speed of oxygen gas molecules at a pressure of $1.01 \times 10^5\,Pa$ when the density is $1.5\,kg\,m^{-3}$.

The Boltzmann constant, E_k, and T

The ideal gas equation $pV = nRT$ where $n = \frac{m}{M}$ (Topic 8.2) is useful for gases with mass m in kg. When you are considering individual molecules it is useful to write the equation in terms of number of molecules, N.

$N = nN_A$ (number of moles × the Avogadro constant)

$$pV = \frac{NRT}{N_A}$$

Using the Boltzmann constant, $k = \frac{R}{N_A} = \frac{8.31\,\text{J}\,\text{mol}^{-1}\,\text{K}^{-1}}{6.02 \times 10^{23}\,\text{mol}^{-1}} = 1.38 \times 10^{-23}\,\text{J}\,\text{K}^{-1}$

$pV = NkT$ (1)

Using the kinetic theory equation: $pV = \frac{1}{3}Nmc_{rms}^2$ (2)

$NkT = \frac{1}{3}Nmc_{rms}^2$ (using equations 1 and 2)

$3kT = mc_{rms}^2$ (3)

So we can also use the Boltzmann constant in the equation for the kinetic energy E_k of a gas.

$E_k = \frac{1}{2}mv^2$ so the average molecular kinetic energy of a gas is:

$E_k = \frac{1}{2}mc_{rms}^2$ where m is the mass of a molecule and c_{rms}^2 is the mean square speed.

$E_k = \frac{3kT}{2}$ (using equation 3)

WORKED EXAMPLE

Find the rms speed of hydrogen ions ($m = 1.67 \times 10^{-27}\,\text{kg}$) at a temperature of $6.0 \times 10^3\,\text{K}$.

$E_k = \frac{1}{2}mc_{rms}^2 = \frac{3kT}{2}$

$c_{rms}^2 = \frac{3kT}{m}$

$c_{rms}^2 = \frac{3(1.38 \times 10^{-23}\,\text{J}\,\text{K}^{-1})(6.0 \times 10^3\,\text{K})}{(1.67 \times 10^{-27}\,\text{kg})} = 1.5 \times 10^8\,\text{m}^2\,\text{s}^{-2}$

$c_{rms} = 1.2 \times 10^4\,\text{m}\,\text{s}^{-1}$

PRACTICE QUESTIONS

3 a Calculate the total kinetic energy of 1 mole of oxygen gas at:

    i 0°C   **ii** 100°C.

b Calculate the mean kinetic energy of an oxygen molecule at:

    i 0°C   **ii** 100°C.

4 The rms speed of a nitrogen molecule is $515\,\text{m}\,\text{s}^{-1}$. Calculate the temperature of the nitrogen gas. (The relative molecular mass of $N_2 = 28$.)

5 A container of hydrogen gas has a pressure of $2.7 \times 10^3\,\text{Pa}$ at 47 K.

a Determine the temperature at which the pressure be $2.4 \times 10^4\,\text{Pa}$.

b If the rms speed at 47 K was $8.0 \times 10^2\,\text{m}\,\text{s}^{-1}$, calculate the rms speed at the new temperature.

8.4 Internal energy

Internal energy – ideal and real

Ideal gas molecules are points with no volume. They can have an elastic collision with another molecule or the walls of the container, but they have no other interactions. This is a useful model, especially for common gases at room temperature and atmospheric pressure. Real gases are not like this.

The internal energy, U, of a system is the energy of its molecules. This is the sum of the randomly distributed kinetic and potential energies of the molecules. $U = E_K + E_P$

For an ideal gas, and for monatomic gases that approximate to ideal, internal energy is the sum of the kinetic energies of the molecules, E_K, due to their movement from one place to another. This is their translational kinetic energy. $U = E_K$

For real molecular gases, E_K includes their translational kinetic energies and their vibrational and rotational kinetic energies as the atoms in the molecule can vibrate and rotate about each other. $U = E_K$

For liquids and solids, there are attractive forces between the atoms or molecules so they have potential energy E_P as well as kinetic energy. $U = E_K + E_P$

Change of phase

Also called a change of state, this is when matter changes between the solid and liquid phases, or between the liquid and gas phases.

Specific latent heat

The thermal energy required to change the state of a material from solid to liquid is called the latent heat of fusion. This is increasing the internal energy by increasing the potential energy of the molecules as the material changes from solid to liquid.

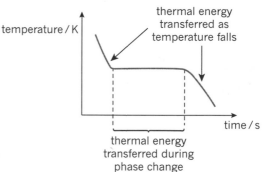

A cooling curve

Similarly, the thermal energy required to change the state from liquid to gas is the latent heat of vapourisation.

The specific latent heat, L, of a material is the amount of thermal energy required to change the phase of 1 kg. It is measured in $J\,kg^{-1}$.

The thermal energy, $\triangle Q$, required to change the phase of a mass m of the material is $\triangle Q = mL$.

To change ice at $0\,^{\circ}C$ to water at $0\,^{\circ}C$: the specific latent heat of fusion $L = 334\,kJ\,kg^{-1}$.

To change water at $100\,^{\circ}C$ to steam at $100\,^{\circ}C$: the specific latent heat of vapourisation $L = 2230\,kJ\,kg^{-1}$.

Specific heat capacity

The thermal energy required to increase the temperature of a material by 1 K is called its heat capacity. The thermal energy increases the average kinetic energy of the molecules, increasing their temperature.

The specific heat capacity, C, of a material is the amount of thermal energy required to raise the temperature of 1 kg of material by 1 K. It is measured in $J\,kg^{-1}\,K^{-1}$

The thermal energy, $\triangle Q$ required by a mass, m, to increase its temperature from θ_1 to θ_2 or by $\triangle\theta = (\theta_2 - \theta_1)$ is:

$$\triangle Q = mC\,\triangle\theta$$

To raise the temperature of water by 1 K: the specific heat capacity of water is $4.2\,\text{kJ}\,\text{kg}^{-1}\text{K}^{-1}$.

WORKED EXAMPLE

The graph shows how the temperature changed when 2.0 kg of copper was heated until it had all melted. Thermal energy was transferred to the copper at a constant rate by a 4.0 kW heater. Assuming all the energy was transferred to the copper, calculate:

a the specific latent heat

b the specific heat capacity of solid copper.

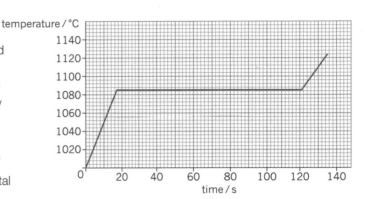

a Time taken for copper to change state = horizontal part of graph = 121 s – 17 s = 104 s
Energy supplied by heater = $\triangle Q = Pt$ and $\triangle Q = mL$
$(4.0 \times 10^3\,\text{W})(104\,\text{s}) = (2.0\,\text{kg})L$

$$L = \frac{(4.0 \times 10^3\,\text{W})(104\,\text{s})}{(2.0\,\text{kg})} = 208\,000\,\text{J}\,\text{kg}^{-1} = 210\,\text{kJ}\,\text{kg}^{-1}\,(2\,\text{s.f.})$$

b Between 0 s and 17 s the temperature of the solid copper increases from 1000 °C to 1084 °C
$\triangle Q = Pt = mC\,\triangle\theta\,(4.0 \times 10^3\,\text{W})(17\,\text{s}) = (2.0\,\text{kg})C\,(1084\,°\text{C} - 1000\,°\text{C})$

$$C = \frac{(4.0 \times 10^3\,\text{W})(17\text{s})}{(2.0\,\text{kg})(84\,\text{K})} = 404\,\text{J}\,\text{kg}^{-1}\text{K}^{-1}$$

> **REMEMBER:** The unit 1 kelvin is the same size as the unit 1 °C. When you are working out temperature differences, you can use either scale, there is no need to convert to K:
> 373 K − 273 K = 100 K = 100 °C − 0 °C = 100 °C

PRACTICE QUESTIONS

1 Using the graph above, calculate the specific heat capacity of liquid copper.

2 45 g of ice at 0 °C is added to a beaker containing 800 g of water at 22 °C. The final temperature of the water and ice is 18 °C. Using the specific heat capacity of water = 4.2 kJ kg⁻¹K⁻¹ calculate the specific latent heat of water. Explain why your answer is less than the actual value.

3 A 550 g block of aluminium at 400 °C is dropped into an insulated beaker containing 750 g of water at 25 °C. Calculate the final temperature of the water and aluminium, assuming no heat is transferred to the beaker and surroundings. The specific heat capacity of aluminium = 897 J kg⁻¹K⁻¹ and of water = 4200 J kg⁻¹K⁻¹.

9 FIELDS

9.1 Gravity

Gravitational force

For two point masses, or two spherical masses of uniform density, M and m, a distance r apart:

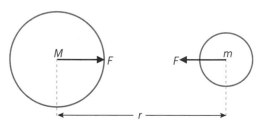

Each mass attracts the other with a force F, which is directly proportional to the masses and inversely proportional to the square of the distance between them.

$F \propto \dfrac{Mm}{r^2}$ This is Newton's law of gravitation and is an example of an inverse square law.

$F = \dfrac{GMm}{r^2}$ where G is the gravitational constant, $G = 6.67 \times 10^{-11}\,\mathrm{N\,m^2\,kg^{-2}}$.

WORKED EXAMPLE

Calculate the gravitational attraction between the Earth and the Moon.
$M = 5.98 \times 10^{24}\,\mathrm{kg}$, $m = 7.35 \times 10^{22}\,\mathrm{kg}$, $r = 3.84 \times 10^8\,\mathrm{m}$

$F = \dfrac{(6.67 \times 10^{-11}\,\mathrm{N\,m^2\,kg^{-2}})(5.98 \times 10^{24}\,\mathrm{kg})(7.35 \times 10^{22}\,\mathrm{kg})}{(3.84 \times 10^8\,\mathrm{m})^2} = 1.99 \times 10^{20}\,\mathrm{N}$

PRACTICE QUESTIONS

1   Calculate the gravitational force between the Sun and the Earth (Sun's mass = $1.99 \times 10^{30}\,\mathrm{kg}$, Earth's mass = $5.98 \times 10^{24}\,\mathrm{kg}$, Earth–Sun distance = $1.50 \times 10^{11}\,\mathrm{m}$).

2   Calculate the gravitational force between two 10.0 kg masses 10.0 m apart.

> **REMEMBER:** The distance is squared. A common mistake is to divide by r and not r^2.

Gravitational field

When a mass experiences a gravitational force it is in a gravitational field. The gravitational field strength, g, is the force per unit mass.

$g = \dfrac{\text{gravitational force}}{\text{mass}}$ $g = \dfrac{F}{m}$

Gravitational field strength is a vector. Its direction is the direction a mass would move in the field. Notice from the diagram of gravitational field that this is *opposite* to the direction of increasing r, so the field strength is negative and its magnitude comes from:

$F = \dfrac{GMm}{r^2}$ and $F = mg$

Gravitational field strength $g = \dfrac{-GM}{r^2}$

Note that the force F on a mass m is then $F = mg$ so F is negative. Some specifications give the magnitude of the force on the formulae sheet and some include the direction: $F = \frac{-GMm}{r^2}$.

(Check which one you must use.)

The gravitational field is radial — the field lines cross at the point in the centre of the mass. To sketch a radial field, place your ruler so that you can draw a line through the central point, but lift your pencil off the paper inside the mass where the equation does not apply.

You must use a ruler, and to make sure the lines are evenly spaced draw the two horizontal and the two vertical lines. If you choose to draw any more, make sure the angles are all equal.

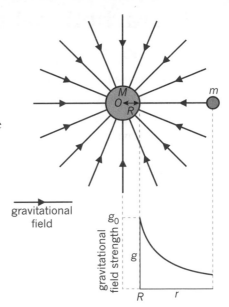

gravitational field

WORKED EXAMPLE

Calculate the gravitational field strength at the surface of the Earth, assuming the mass of the Earth is concentrated at a point at the centre (Earth's mass $= 5.98 \times 10^{24}$ kg, radius $= 6.38 \times 10^6$ m).

$$g = \frac{-GM}{r^2} = \frac{-(6.67 \times 10^{-11}\,\text{N}\,\text{m}^2\,\text{kg}^{-2})(5.98 \times 10^{24}\,\text{kg})}{(6.38 \times 10^6\,\text{m})^2} = -9.80\,\text{N}\,\text{kg}^{-1}$$

(Notice that this is close to the accepted value of $-9.81\,\text{N}\,\text{Kg}^{-1}$.)

PRACTICE QUESTIONS

3   Calculate the gravitational field strength at the surface of Jupiter, assuming the mass of Jupiter is concentrated at a point at the centre. (Jupiter's mass $= 1.90 \times 10^{27}$ kg, radius $= 7.14 \times 10^7$ m.)

4   Using the fact that the gravitational field strength on the Moon is one sixth that on the Earth, and the Moon's radius is 1.74×10^6 m, calculate a value for the mass of the Moon.

STRETCH YOURSELF!

a   Calculate a value for the gravitational field strength of the Earth 1000 m above the surface of the Earth.

(Earth's mass $= 5.98 \times 10^{24}$ kg, radius $= 6.38 \times 10^6$ m.)

b   Use your answer to suggest whether it is appropriate to use a constant value for the Earth's gravitational field of $g = 9.81\,\text{N}\,\text{kg}^{-1}$ in calculations.

c   Show that the distance x from the Earth where the gravitational field strengths of the Earth and Moon are equal can be represented by the equation $x^2 - 7.8 \times 10^8 x + 1.5 \times 10^{17} = 0$. This equation has two solutions: $x = 4.4 \times 10^8$ m and 3.4×10^8 m. Explain which is correct. (Earth's mass $= 5.98 \times 10^{24}$ kg, Moon's mass $= 7.35 \times 10^{22}$ kg, Earth–Moon distance $= 3.84 \times 10^8$ m.)

9.2 Gravitational potential

Work done in a gravitational field

Two masses have zero gravitational potential energy when they are an infinite distance apart because their masses have no effect on each other.

As a rocket carries you away from the Earth's surface, the gravitational field gets weaker the greater the distance you travel from the Earth. The force from the rocket engines is doing work on you and your potential energy is increasing. Only when you reach an infinite distance from Earth is the gravitational force pulling you back to Earth equal to zero. As this is defined as the zero of potential energy and your potential energy has increased, it follows that on the surface of the Earth your potential energy was negative.

Note that all the calculations you have done involving potential energy as objects are raised and lowered near the surface of the Earth have been increases and decreases of potential energy. It is not necessary to know where the zero is to calculate increases and decreases.

The work done, $\triangle W$, by the gravitational force to bring a mass m from infinity to a distance r from a mass M is:

$$\triangle W = \frac{-GMm}{r}$$ This is also the gravitational potential energy.

WORKED EXAMPLE

Calculate **a** the potential energy of a 600 kg mass on the Earth's surface and **b** its increase in potential energy when lifted 500 km above the surface. (Earth's mass = 5.98×10^{24} kg, radius = 6.38×10^6 m.)

a   At the surface:

$$E_P = \frac{-GMm}{r} = \frac{-(6.67 \times 10^{-11}\,Nm^{-2}\,kg^{-2})(5.98 \times 10^{24}\,kg)(600\,kg)}{(6.38 \times 10^6\,m)} = -3.75 \times 10^{10}\,J$$

b   500 km above the surface, the height has changed so replace r in the equation with the new height (multiply the answer to part **a** by the old height and divide by the new height):

$$E_P = \frac{(-3.75 \times 10^{10}\,J)(6.38 \times 10^6\,m)}{(6.38 \times 10^6\,m + 500 \times 10^3\,m)} = -3.48 \times 10^{10}\,J$$

PRACTICE QUESTIONS

1   Calculate the potential energy of an 85 kg mass at the surface of Jupiter. (Jupiter's mass = 1.9×10^{27} kg, radius = 7.0×10^7 m.)

2   Calculate the minimum work done on a satellite of mass 1100 kg when it is launched from the surface of the Earth into an orbit 36 000 km above the surface. (Earth's radius = 6400 km, mass = 6.0×10^{24} kg.)

Gravitational potential

The gravitational potential, V, at a point in a gravitational field is the gravitational potential energy, per unit mass, of a small point mass. It is measured in $J\,kg^{-1}$.

$$V = \frac{-GM}{r}$$

Gravitational potential at a point in a gravitational field is defined as the work done by the gravitational field to bring a mass from infinity to that point. It is negative because work is done by the field. This results in a lower gravitational potential energy. (When work is done against the field to raise a mass, gravitational potential energy increases and the mass is then at a point with a higher gravitational potential.).

REMEMBER:
Gravitational potential is always negative. There are no repulsive gravitational forces.

Gravitational potential at a distance r from the Earth

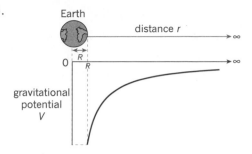

WORKED EXAMPLE

Calculate **a** the gravitational potential at the surface of the Earth and **b** the escape velocity required for a mass on the surface.
(Earth radius = 6 400 km mass = 6.0×10^{24} kg)

a $V = \dfrac{-GM}{R} = \dfrac{-(6.67 \times 10^{-11}\,\text{Nm}^{-2}\,\text{kg}^{-2})(6.0 \times 10^{24}\,\text{kg})}{(6.4 \times 10^{6}\,\text{m})} = -6.3 \times 10^{7}\,\text{J}\,\text{kg}^{-1}$

The kinetic energy + potential energy at the surface of the Earth = $\frac{1}{2}mv^2 - \dfrac{GMm}{r}$

As the object travels away, it loses kinetic energy and gains gravitational potential energy. When the KE = 0, if it has not escaped the gravitational field, it will fall back to Earth.

b A mass m will have just enough energy to escape when $\frac{1}{2}mv_{esc}^2 = \dfrac{GMm}{r}$, which can be rearranged to:

$\frac{1}{2}v_{esc}^2 = \dfrac{GM}{R}$

$v_{esc}^2 = 2(-V)$

so $v_{esc} = \sqrt{(-2V)} = \sqrt{(-2(-6.3 \times 10^{7}\,\text{J}\,\text{kg}^{-1}))} = \sqrt{(12.6 \times 10^{6}\,\text{J}\,\text{kg}^{-1})} = 1.1 \times 10^{4}\,\text{m}\,\text{s}^{-1}$ *or* 11 km s^{-1}

PRACTICE QUESTIONS

Use Earth's mass = 5.98×10^{24} kg, radius = 6.38×10^{6} m.

3 Calculate the gravitational potential at a point:

  a 6380 km above the Earth's surface.

  b at the surface of Mars (mass = 0.107 × Earth's mass, radius = 0.532 × Earth's radius)

4 Use your answer to Question **3b** to calculate the escape velocity from Mars.

STRETCH YOURSELF!

The Swarzschild radius is the radius that a black hole would have if the escape velocity was the speed of light. Calculate the Swarzschild radius of the supermassive black hole at the centre of the Milky Way (mass = 8.2×10^{36} kg).

9.3 Orbital motion

Centripetal force

To keep an object moving in a circular path requires a centripetal force (see Topic 7.1). The planets move in elliptical orbits, but you can investigate the motion of satellites moving in circular orbits and in most cases treat the planetary orbits as approximate circles.

The gravitational force provides the resultant centripetal force to keep the planet in orbit:

$$\frac{GmM}{r^2} = \frac{mv^2}{r} \quad \textbf{(1)}$$

All of the equations for circular motion can then be used. For example:

$$T = \frac{2\pi}{\omega} \quad v = r\omega \quad a = r\omega^2$$

WORKED EXAMPLE

Calculate the force needed to keep the Earth in orbit around the Sun.
(Earth's mass $= 5.98 \times 10^{24}$ kg, radius of orbit $= 1.50 \times 10^{11}$ m.)

The time for the Earth to orbit the Sun $= 365.3$ days $= 365.3 \times 24 \times 60 \times 60$ s $= 3.16 \times 10^7$ s

$$F = mr\omega^2 \quad \omega = \frac{2\pi}{T} = \frac{2\pi}{(3.16 \times 10^7 \,\text{s})} = 1.99 \times 10^{-7} \,\text{rad s}^{-1}$$

$$F = (5.98 \times 10^{24} \,\text{kg})(1.50 \times 10^{11} \,\text{m})(1.99 \times 10^{-7} \,\text{rad s}^{-1})^2 = 3.55 \times 10^{22} \,\text{N}$$

Compare this to your answer to Question 2 from Topic 9.1.

PRACTICE QUESTION

1   A geostationary orbit has a period of 24 hours so that the satellite remains at the same point above the Earth's surface as the Earth rotates. Calculate the radius of the geostationary orbit. (Earth's mass $= 5.98 \times 10^{24}$ kg.)

Power laws

From equation **(1)** above: $v^2 = \frac{Gm}{r}$.

Using the equation for the period of the orbital motion: $T = \frac{2\pi r}{v}$ gives: $\frac{(2\pi r)^2}{T^2} = \frac{Gm}{r}$.

$T^2 = \frac{4\pi^2 r^3}{GM}$ or $T = \sqrt{\frac{4\pi^2 r^3}{GM}}$, which can be written: $T = \left(\frac{2\pi}{\sqrt{GM}}\right) r^{3/2}$ **(2)**.

This is an example of a power law. In this case, the period squared is directly proportional to the radius cubed:

$$T^2 \propto r^3$$

Plotting T against r will give us a curve, from which it is not easy to tell whether this relationship is correct.

Power laws can be tested by plotting a log–log graph, because this gives a straight line, and the gradient will tell us the power.

In this case you can plot $\log_{10}(T)$ against $\log_{10}(r)$ (you say 'log to base ten of T against log to base ten of r'). You could also use natural logs, which are written $\log_e$ or $\ln$, see Topic 10.3.

Logarithms

$10^2 \times 10^3 = 10^5$ When the numbers are multiplied the powers are added.

Logarithms are the powers of the numbers, so $\log_{10}(10^2) = 2$ and $\log_{10}(10^3) = 3$.

To multiply the two numbers you add the logarithms: $2 + 3 = 5$, and then the final answer is the antilog$_{10}(5) = 10^5$.

REMEMBER: Logs – the basics

Example	Log₁₀
$10 = 10^1$	$\log_{10}(10) = 1$
$0.1 = \frac{1}{10} = 10^{-1}$	$\log_{10}(0.1) = -1$
$1 = 10^{1-1} = 10^0$	$\log_{10}(1) = 0$
0	$\log_{10}(0) = -\infty$
X^A	$\log_{10}(X^A) = A\log_{10}(X)$
$X^A\, Y^B$	$\log_{10}(X^A\, Y^B) = A\log_{10}(X) + B\log_{10}(Y)$
$\frac{X^A}{Y^B} = X^A\, Y^{-B}$	$\log_{10}\left(\frac{X^A}{Y^B}\right) = A\log_{10}(X) - B\log_{10}(Y)$

Taking $\log_{10}$ of equation **(2)**:

$$\log_{10}(T) = \log_{10}\frac{(4\pi)^{\frac{1}{2}}}{(GM)^{\frac{1}{2}}} + \log_{10}(r^{\frac{3}{2}})$$

$$\log_{10}(T) = \log_{10}\frac{(4\pi^2)^{\frac{1}{2}}}{(GM)^{\frac{1}{2}}} + \frac{3}{2}\log_{10}(r) \quad \text{Plotting } \log_{10}(T) \text{ against } \log_{10}(r) \text{ gives a straight line.}$$

$$y = \qquad c \qquad + \qquad mx$$

with gradient $m = \dfrac{3}{2} = 1.5$

WORKED EXAMPLE

You can test equation **(2)** for the planets:

NOTE:
In this example you are just checking the gradient. If you need to find the constant, c, remember that you will need to choose your axes so that you can read off the y-intercept. Using the mass of the Sun to calculate the value of c gives about -9.3, so your y-axis has to go down to about -10.

Gradient $m = \dfrac{3}{2} = 1.5$

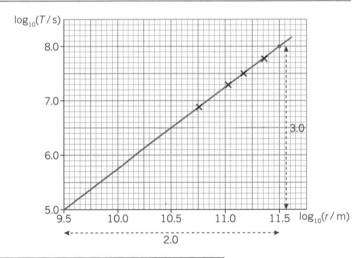

Planet	Period T/days	Period T/s	Log₁₀ (T/s)	Radius of orbit r/m	Log₁₀ (r/m)
Mercury	88	7.60×10^6	6.88	0.587×10^{11} m	10.76
Venus	225	1.94×10^7	7.29	1.08×10^{11} m	11.03
Earth	365	3.16×10^7	7.50	1.50×10^{11} m	11.17
Mars	687	5.94×10^7	7.77	2.28×10^{11} m	11.36

PRACTICE QUESTION

2 Plot a graph for all eight planets. Use this graph:

  a to determine whether $T^2 \propto r^3$

  b to calculate a value for the mass of the Sun.
    (Hint: You will need to use the value of c.)

  Use the data for the four planets above, and also:

Planet	Period T/days	Radius of orbit r/m
Jupiter	4332	7.78×10^{11} m
Saturn	10762	1.43×10^{12} m
Uranus	30693	2.87×10^{12} m
Neptune	60201	4.50×10^{12} m

9.4 Electric fields

Coulomb's law

Electric charges can be positive or negative, so the electric force between them is attractive for opposite charges and repulsive for like charges. Electric charge is measured in coulombs (C).

For two point charges, or two spherical charges, Q and q, a distance r apart:

Electric force

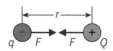

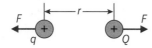

The force F is directly proportional to the charges and inversely proportional to the square of the distance between them:

$F \propto \dfrac{Qq}{r^2}$ This is Coulomb's law and is an example of an inverse square law.

$F = \dfrac{kQq}{r^2}$ k depends on the material between the charges.

For a vacuum, $k = 8.99 \times 10^9\,\mathrm{N\,m^2\,C^{-2}}$.

k is effectively the same for air. Another way of expressing Coulomb's law is:

$F = \dfrac{Qq}{4\pi\varepsilon_0 r^2}$

Where ε_0 is the permittivity of free space, $\varepsilon_0 = 8.85 \times 10^{-12}\,\mathrm{C^2\,N^{-1}\,m^{-2}}$.

$k = \dfrac{1}{4\pi\varepsilon_0}$

For like charges, the force F is positive; for unlike charges, the force is negative.

WORKED EXAMPLE

Calculate the force on an electron in the hydrogen atom when the distance from the nucleus = $5.3 \times 10^{11}\,\mathrm{m}$.

$$F = \frac{(+1.60 \times 10^{-19}\,\mathrm{C})(-1.60 \times 10^{-19}\,\mathrm{C})}{4\pi(8.85 \times 10^{-12}\,\mathrm{C^2\,N^{-1}\,m^{-2}})(5.3 \times 10^{-11}\,\mathrm{m})^2} = -8.2 \times 10^{-8}\,\mathrm{N}$$

PRACTICE QUESTION

1 A charge of $3.0 \times 10^{-9}\,\mathrm{C}$ is $0.3\,\mathrm{m}$ from a charge of $6.0 \times 10^{-9}\,\mathrm{C}$. Calculate the repulsive force between them.

> **REMEMBER:** The distance is squared. A common mistake is to divide by r and not r^2.

Electric field – point charge

An electric charge creates an electric field around it. The electric field strength, E, at a point in space is the force per unit positive charge experienced by a point charge at that point in space.

$E = \dfrac{\text{electric force}}{\text{charge } q}$ $E = \dfrac{F}{q}$

Electric field strength is a vector. Its direction is the direction a positive charge would move in the field. The magnitude comes from:

Electric field strength $E = \dfrac{Q}{4\pi\varepsilon_0 r^2}$ and $F = Eq$

WORKED EXAMPLE

Calculate the magnitude of the electric field 5.0 nm from a carbon nucleus. (Consider the nucleus as a point charge.)

$$E = \frac{Q}{4\pi\varepsilon_0 r^2} = \frac{12(+1.60 \times 10^{-19}\,\text{C})}{4\pi(8.85 \times 10^{-12}\,\text{C}^2\,\text{N}^{-1}\,\text{m}^{-2})(5.0 \times 10^{-9}\,\text{m})^2} = 6.9 \times 10^{8}\,\text{N}\,\text{C}^{-1}$$

Electric field – parallel plates

This diagram shows the electric field strength between two parallel plates a distance d apart with a pd of V between them.

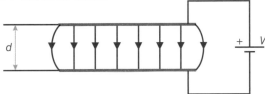

The electric field E is uniform, so there is a constant potential gradient between the two plates: $E = \dfrac{V}{d}$.

Electric field can be expressed in units of $\text{V}\,\text{m}^{-1}$ or $\text{N}\,\text{C}^{-1}$.

A charged particle between the two plates will experience a force towards the oppositely charged plate. For an electron this will be $F = Eq = Ee = \dfrac{Ve}{d}$.

WORKED EXAMPLE

Two parallel plates are 2.5 cm apart in a vacuum. The pd between them is 1.5 kV. The electric field strength is:

$$E = \frac{V}{d} = \frac{(1.5 \times 10^3\,\text{V})}{(2.5 \times 10^{-2}\,\text{m})} = 6.0 \times 10^4\,\text{V}\,\text{m}^{-1}$$

PRACTICE QUESTIONS

2 Calculate the magnitude and the direction of the electric field 10.0 cm from an electron.

3 At a point X the electric force on a point charge of $6.0 \times 10^{-9}\,\text{C}$ is $4.5 \times 10^{-3}\,\text{N}$. Calculate the magnitude of the electric field at X.

4 a Calculate the electric field between two parallel plates a distance 1.8 cm apart in a vacuum when the pd between the plates is 3.5 kV.

   b Calculate the force on an electron between the two plates.

5 Show that the units $\text{V}\,\text{m}^{-1}$ are the same as the units $\text{N}\,\text{C}^{-1}$.

9.5 Electric potential

Electric potential energy

Two charges have zero potential energy when they are an infinite distance apart because their charges have no effect on each other. This is a similar case to gravitational potential energy (see Topic 9.2).

The work done, $\triangle W$, by the electric (or electrostatic) force to bring a charge $-q$ from infinity to a distance r from a charge $+Q$ is:

$\triangle W = -\dfrac{kQq}{r}$ This is also the electric potential energy.

This is the same for a charge $+q$ brought from infinity to a distance r from a charge $-Q$.

Unlike the force of gravity, which is always attractive, the electric force can also be a repulsive force between two positive charges or two negative charges.

Two positive charges repel, so to bring a charge $+q$ from infinity to a distance r from a charge $+Q$ you would need to do work on the charge $+q$.

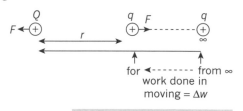

The electric force, F, is pushing the charges apart, so an opposite F is needed to move them closer. They will have more potential energy than they did at infinity, so if they can move, the electric force will move them apart.

$\triangle W = \dfrac{kQq}{r}$ This is also the electric potential energy, E_p.

This would be the same for charges $-Q$ and $-q$.

> **REMEMBER:**
> $k = \dfrac{1}{4\pi\varepsilon_0}$ (see Topic 9.4)
> $k = 8.99 \times 10^9$ N m^2 C^{-2}. It is not the Boltzmann constant (this is a common mistake).

WORKED EXAMPLE

The electric potential energy of two alpha particles is 1.35×10^{-13} J. Calculate their distance apart.

$(1.35 \times 10^{-13}\ \text{J}) = \dfrac{(8.99 \times 10^9\,\text{N}\,\text{m}^2\ \text{C}^{-2})(2 \times 1.60 \times 10^{-19}\ \text{C})^2}{r} = \dfrac{9.206 \times 10^{-28}\ \text{J}}{r}$

$r = \dfrac{9.206 \times 10^{-28}}{1.35 \times 10^{-13}} = 6.82 \times 10^{-15}$ m

---

PRACTICE QUESTIONS

1   A chlorine ion (Cl$^-$) and an oxygen ion (O^{2-}) are 1.20 nm apart. Calculate their electric potential energy.

2   In a hydrogen atom, the electron is 5.29×10^{-11} m from the proton. Calculate the work done when the electron is moved away from the proton, to infinity. (In real life, this is a distance far enough away to be a free electron.)

Electric potential

The electric potential, V, at a point in an electric field is the electric potential energy, per unit charge, of a small positive point charge. It is measured in J C^{-1} or volts V.

$$V = \frac{Q}{4\pi\varepsilon_0 r} \quad \textbf{(1)} \qquad \text{or} \qquad V = \frac{kQ}{r}$$

Electric potential at a point in an electric field is defined as the work done by the electric field to bring a positive charge from infinity to that point. It is positive because work is done against the field. This results in a higher electric potential energy. (When work is done by the field to attract a charge, electric potential energy decreases and the charge is then at a point with a lower electric potential.)

This gives an alternative equation for the work done, $\triangle W$, in moving a charge, Q, through a potential difference $\triangle V$. This is also the electric potential energy, E_P.

$$\triangle W = E_P = Q\triangle V$$

Around a point charge, Q, the electric field $E = \frac{Q}{4\pi\varepsilon_0 r^2}$ (see Topic 9.4) so from

(1) above: $E = \frac{\triangle V}{\triangle r}$

The graphs below show how the electric force, field, and potential each vary with the distance r from a point charge Q.

The electric field, E, is equal to the gradient of the graph of potential, V, against r.

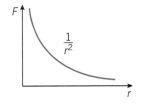
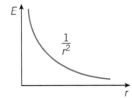
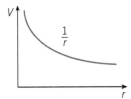

WORKED EXAMPLES

1 Calculate the electric potential at a point P that is 3.25 m from a 475 nC point charge.

$$V = \frac{kQ}{R} = \frac{(8.99 \times 10^9 \,\text{N m}^2 \,\text{C}^{-2})(475 \times 10^{-9}\text{C})}{(3.25\,\text{m})} = 1.31 \times 10^3 \,\text{V}$$

2 A charged metal sphere of diameter 10.0 cm is insulated from its surroundings.
It is charged from a +1.2 kV supply. Calculate the charge on its surface.

The potential on the surface = +1.2 kV. The sphere can be considered as a point charge at its centre.

$$(1.2 \times 10^3 \,\text{V}) = \frac{(8.99 \times 10^9 \,\text{N m}^2 \,\text{C}^{-2})Q}{(5.0 \times 10^{-2}\,\text{m})}$$

$$Q = \frac{(1.2 \times 10^3 \,\text{V})(5.0 \times 10^{-2}\,\text{m})}{(8.99 \times 10^9 \,\text{N m}^2 \,\text{C}^{-2})} = 6.67 \times 10^{-9} \,\text{C}$$

PRACTICE QUESTION

3 The electric potential 255 pm from a charged particle is 17 V.

    a Calculate the charge.

    b Determine the magnitude of the electric field at this point.

9.6 Magnetic fields

Magnetic force and flux density

When a current-carrying conductor is perpendicular to a uniform magnetic field, the field exerts a force on the conductor. The direction of the force is given by Fleming's left-hand rule: thumb = force, forefinger = field, and index finger = current (direction of movement of a positive charge).

The magnitude of the force, F, depends on the current, I, the length of the conductor, l, and the strength of the magnetic field, B.

This effect defines the magnetic field strength, called the magnetic flux density, B.

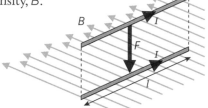

$$F = BIl$$

B is a vector and the unit is the tesla (T).

If the conductor is not at right angles to the field, the effect is reduced. When θ is the angle between the conductor and the field:

$$F = BIl\sin\theta$$

When the conductor is parallel to the field $\theta = 0$ and there is no force on the conductor.

WORKED EXAMPLE

A straight conductor length 50 cm is at right angles to a magnetic field of flux density 0.90 T. If the force on it is 9.0 N, calculate the current through it.

$F = BIl\sin\theta$ where $F = 9.0\,\text{N}$, $B = 0.90\,\text{T}$, $I = ?$, $l = 50\,\text{cm}$, $\theta = 90°$

$$I = \frac{(9.0\,\text{N})}{(0.90\,\text{T})(50 \times 10^{-2}\,\text{m})} = 20\,\text{A}$$

PRACTICE QUESTION

1   Calculate the force on a 12 cm straight wire carrying a current of 360 mA perpendicular to a field with magnetic flux density of $2.0 \times 10^{-3}\,\text{T}$.

Force on a moving charge

When a length of wire l with cross-sectional area A contains n charges per unit volume, the total number of charges in the section of wire $N = nAl$ (n is called the number density of the charges).

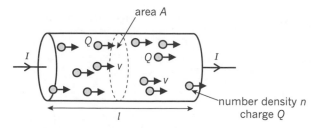

If the wire is perpendicular to a magnetic field, the total force F_T on the wire due to the magnetic field on all the charges $F_T = BIl$.

The force on one charge in the wire: $F = \dfrac{BIl}{nAl}$

The current in the wire $I = nAve$ if the charge carriers are electrons (see Topic 5.2). If they are particles with charge Q, for example ions, then $I = nAvQ$.

$F = \dfrac{BnAvQl}{nAl} = BQv$ where v is the velocity of the particle.

If the particle is travelling at an angle θ to the field the equation becomes:

$F = BQv \sin \theta$

If the force is at right angles to the velocity of the particle, the particle will move in a circular path (note that a $\times$ means a magnetic field moving into the page and a $\bullet$ means a magnetic field moving out of the page):

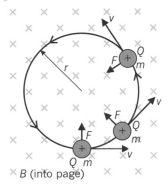

$\times B$ (into page)

The resultant force on the particle from the magnetic field provides the centripetal force (see Topic 7.1).

$BQv = \dfrac{mv^2}{r}$ or $r = \dfrac{mv}{BQ}$ Notice that the radius is proportional to the speed and increasing B decreases r.

When the particle is an electron the force is Bev and the radius is $r = \dfrac{mv}{Be}$.

This is the principle behind the cyclotron for accelerating charged particles and also for the mass spectrometer for separating ions of different mass.

 ## WORKED EXAMPLE

Two singly ionised atoms leave a slit travelling with the same velocity $v = 1.0 \times 10^6 \, \text{m s}^{-1}$ and enter a magnetic field of 0.10 T. The first ion is a hydrogen ion of mass 1.67×10^{-27} kg. The second ion is a deuterium ion of mass 3.34×10^{-27} kg. They both move in a circular path. Find their separation after they have both completed a semicircle.

$r = \dfrac{mv}{BQ}$ 1st ion: $r_1 = \dfrac{(1.67 \times 10^{-27} \, \text{kg})(1.0 \times 10^6 \, \text{m s}^{-1})}{(0.10 \, \text{T})(1.60 \times 10^{-19} \, \text{C})} = 0.10 \, \text{m}$

2nd ion: $r_2 = \dfrac{(3.34 \times 10^{-27} \, \text{kg})(1.0 \times 10^6 \, \text{m s}^{-1})}{(0.10 \, \text{T})(1.60 \times 10^{-19} \, \text{C})} = 0.21 \, \text{m}$

Separation $= (2r_2 - 2r_1) = 2 \times (0.21 \, \text{m}) - 2 \times (0.10 \, \text{m}) = 0.22 \, \text{m}$

 ## PRACTICE QUESTIONS

2 a A proton is moving in a circle of radius 14 cm in a uniform magnetic field of 0.35 T, perpendicular to its direction of movement. Calculate the velocity of the proton.

b If an electron moves with the same velocity perpendicular to the same field, determine the radius of its path.

3 A proton with kinetic energy $E_K = 2.5 \times 10^{-14}$ J moves perpendicularly to a magnetic field of 0.27 T in a cloud chamber. Calculate the radius of its track.

9.7 Electromagnetic induction

Magnetic flux

The magnetic flux, ϕ, is the magnetic flux density B multiplied by the projection of the area A that the flux is passing through onto a surface at right angles to the flux. For example, the flux ϕ through a coil of cross-section A at an angle θ to a magnetic field B:

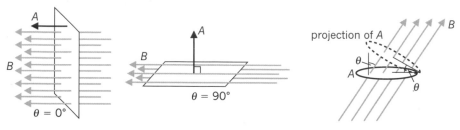

$\phi = BA \cos \theta$

Magnetic flux is measured in webers (Wb).

Magnetic flux linkage

When a coil of wire has a number of turns N, the flux passes through the area A of each loop, or turn, so the magnetic flux linkage is equal to the magnetic flux through the coil $\times$ the number of turns:

$N\phi = BAN \cos \theta$ The units are weber turns (Wb turns).

WORKED EXAMPLE

The Earth's magnetic field strength is 5.0×10^{-5} T. Find the flux linkage through a coil of area $18\,\text{cm}^{-2}$ and nine turns when the plane of the coil is perpendicular to the field ($\theta = 0°$).

$N\phi = BAN \cos \theta = (5.0 \times 10^{-5}\,\text{T})(18 \times 10^{-4}\,\text{m}^2) \times 9 \times 1$

$N\phi = 81 \times 10^{-8}\,\text{Wb turns}$

When the plane of the coil is at an angle of 30° to the field:

$N\phi = BAN \cos \theta = (5.0 \times 10^{-5}\,\text{T})(18 \times 10^{-4}\,\text{m}^2) \times 9 \cos 60°$

$N\phi = 4.1 \times 10^{-7}\,\text{Wb turns}$

PRACTICE QUESTIONS

1 A magnetic field of strength 0.25 T is perpendicular to a circular loop of wire of radius 15 cm. Calculate the flux though the area.

2 A coil of area $100\,\text{cm}^2$ has 50 turns. It is placed so the plane of the coil is at an angle of 30° to a uniform magnetic field of 7.5×10^{-3} T. Calculate the magnetic flux linkage.

Induced emf

When a conductor moves through a magnetic field, according to Faraday's law of electromagnetic induction, there is an induced emf between the ends of the conductor that is proportional to the rate at which the flux is cut:

Induced emf $\propto \dfrac{\text{flux cut}}{\text{time taken}}$ $\varepsilon \propto \dfrac{\Delta \phi}{\Delta t}$

For a coil with N turns:

$$\varepsilon = \frac{N\Delta\phi}{\Delta t}$$

Lenz's law states that the induced emf is always in a direction that opposes the change producing it. On some specifications this is combined with Faraday's law to give the equation:

$$\varepsilon = -\frac{N\Delta\phi}{\Delta t}$$

The units are $\text{Wb s}^{-1} = \text{T m}^2\text{s}^{-1}$ From $F = BIl$ you can see $1\,\text{T} = 1\,\text{NA}^{-1}\text{m}^{-1}$ so $1\,\text{Wb s}^{-1} = 1\,\text{NA}^{-1}\text{m}^{-1}\text{m}^2\text{s}^{-1} = 1\,\text{N m A}^{-1}\text{s}^{-1} = 1\,\text{J C}^{-1} = 1\,\text{V}$

WORKED EXAMPLE

A straight conductor of length 20 cm moves at right angles to a magnetic field of 5.0×10^{-4} T. If it travels 1.2 m in 4.0 s, determine the induced emf between its ends.

The conductor sweeps out an area $A = (20 \times 10^{-2}\,\text{m})(1.2\,\text{m}) = 24 \times 10^{-2}\,\text{m}^2$

$\phi = BA \cos\theta = (5.0 \times 10^{-4}\,\text{T})(24 \times 10^{-2}\,\text{m}^2) \times 1 = 1.2 \times 10^{-4}\,\text{Wb}$

$$\varepsilon = -\frac{N\Delta\phi}{\Delta t} = -\frac{1 \times (1.2 \times 10^{-4}\,\text{Wb})}{(4.0\,\text{s})} = 3.0 \times 10^{-5}\,\text{V}$$

WORKED EXAMPLE

A 500 turn coil of cross-sectional area $17.7 \times 10^{-3}\,\text{m}^2$ has its axis parallel to the Earth's magnetic field. It is turned so that its axis is perpendicular to the Earth's magnetic field in 2.77 ms. The emf induced in the coil is 0.166 V. Calculate the Earth's magnetic field.

$A \cos\theta$ changes from $(17.7 \times 10^{-3}\,\text{m}^2)$ to 0 in 2.77 ms because θ changes from 0° to 90°.

$\Delta\phi = B\Delta(A \cos\theta) = B(17.7 \times 10^{-3}\,\text{m}^2)$

$$\varepsilon = -\frac{N\Delta\phi}{\Delta t}$$

$$(0.166\,\text{V}) = \frac{-(500) \times B(17.7 \times 10^{-3}\,\text{m}^2)}{(2.77 \times 10^{-3}\,\text{s})}$$

$$B = \frac{-(0.166\,\text{V})(2.77 \times 10^{-3}\,\text{s})}{(8.85\,\text{m}^2)} = 5.2 \times 10^{-5}\,\text{T}$$

PRACTICE QUESTIONS

3   An aircraft with metal wings that span 42 m flies horizontally at a speed of 900 km h^{-1}. The Earth's magnetic field is 8.77×10^{-6} T at an angle of 64° to the horizontal. Calculate the emf induced between the ends of the wings.

4   A coil of 50 turns of wire and area 45 cm^2 is placed so that a magnetic field of 1.2×10^{-2} T enters the plane of the coil at right angles. Find the emf induced in the coil when the field:

   a   is reduced to zero in 0.050 s

   b   is increased to 1.6×10^{-2} T in 0.10 s

   c   is reversed in 0.20 s.

1 The diagram below shows a model car travelling on a circular frictionless track of radius 1.5 m at constant speed of 3.2 m s^{-1}. The mass of the car is 64 g.

    a Calculate the centripetal force on the car.

    b Calculate the banking angle required to keep the car on the track.

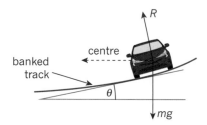

2 A mass of 0.25 kg on a string 0.95 m long forms a simple pendulum. It is set in motion by being struck by a horizontal force so that it oscillates with an angular displacement of 4°. Calculate:

    a the maximum vertical displacement of the mass from the initial position

    b the energy transferred to the mass by the force

    c the maximum speed of the mass.

3 The graph below shows how the pressure of a fixed mass of an ideal gas varies when the volume is changed. The experiment was done at two different temperatures.

Use data from the graph and a suitable equation to calculate:

    a the volume of the gas at Y

    b the pressure of the gas at Z.

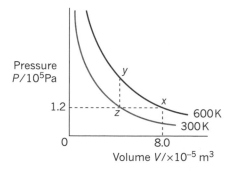

4   a Derive an equation for the speed of the molecules in an ideal gas at temperature T.

    b Use the expression to calculate the speed of neon atoms at 24 °C.

    c Calculate the total internal energy of 35 g of neon at 56 °C.

    Neon can be assumed to be an ideal gas.
    Relative atomic mass neon = 20.

5 A satellite of mass 100 kg orbits the Earth at a height of 2.50×10^6 m. Calculate, for the satellite:

  a the magnitude of the gravitational force

  b the magnitude of the gravitational field

  c the gravitational potential at this height

  d the potential energy.

  Mass of Earth $= 5.98 \times 10^{24}$ kg, radius of Earth $= 6.38 \times 10^6$ m.

6 **a** Calculate the potential at point X, 2.0 mm from a proton.

  b Determine the potential difference between point X and a point Y, 2.5 mm from the proton.

  c Calculate how much work needs to be done to move an electron from point X to point Y.

7 A positive ion with a single charge has a speed of 5.2×10^5 ms^{-1}. It moves in a circular path of radius 7.3 m due to a magnetic field of 1.5 T.

  Calculate the mass of the ion.

8 Electrons moving perpendicularly to a magnetic field of 1.20×10^{-3} T make 3.36×10^7 circular orbits per second.

  Calculate a value for the charge to mass ratio of an electron, e/m.

9 A circular flat coil of radius 4.5 cm has 200 turns of wire. The total resistance of the coil is 3.3 Ω. A uniform magnetic field is applied at right angles to the coil. The field changes uniformly from 0.10 T to 0.75 T in 0.50 s. Calculate:

  a the induced emf in the coil whilst the magnetic field is changing

  b the induced current.

10 A shower has a 10.5 kW electric heater and delivers 12 litres of water per minute. Before it is heated the water temperature is 16 °C. Assuming all the heat is transferred to the water, calculate the temperature of the water in the shower.

  (Specific heat capacity of water = 4200 J kg^{-1} K^{-1})

11 A copper container of mass 120 g contains 180 g of water and a 10 g ice cube. They are all at a temperature of 0 °C in equilibrium with the surroundings. A block of aluminium of temperature 96 °C is added to the container and the final equilibrium temperature is 4 °C.

  Calculate the mass of the aluminium.

  State an assumption made in your calculation.

  Specific heat capacities: copper = 387 J kg^{-1} K^{-1}, water = 4200 J kg^{-1} K^{-1}, aluminium = 900 J kg^{-1} K^{-1}.

  Specific latent heat of fusion for water = 334 000 J kg^{-1}.

10 CAPACITORS

10.1 Capacitors in circuits

Capacitors

Capacitors store energy by separating charge. (They are often said to 'store' charge, but there is a charge of $+Q$ on one plate and $-Q$ on the other so no net charge is stored.)

A capacitor is charged by connecting a pd across it. The greater the pd the greater the charge: $Q \propto V$.

The capacitance C of a capacitor is the charge per unit pd V: $C = \dfrac{Q}{V}$.

Capacitance is measured in farads, F, where $1 \, \text{F} = \text{C} \, \text{V}^{-1}$.

Capacitors in series

When two capacitors, C_1 and C_2, are connected in series in a circuit:

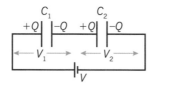

They separate the same charge Q as a single capacitor C.

$$V = \frac{Q}{C} \quad V_1 = \frac{Q}{C_1} \quad V_2 = \frac{Q}{C_2}$$

The pd across both: $V = V_1 + V_2$

So, $\dfrac{Q}{C} = \dfrac{Q}{C_1} + \dfrac{Q}{C_2}$

So they have the same capacitance as a single capacitor, C, where:

$$\frac{1}{C} = \frac{1}{C_1} + \frac{1}{C_2}$$

Capacitors in parallel

When two capacitors, C_1 and C_2, are connected in parallel in a circuit, they have the same pd across them: $Q = CV \quad Q_1 = C_1V \quad Q_2 = C_2V$

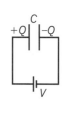 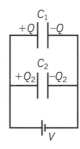

They each separate charge, so $Q = Q_1 + Q_2$

So, $CV = C_1V + C_2V$

So they have the same capacitance as a single capacitor C where:
$C = C_1 + C_2$

 WORKED EXAMPLE

The charge on a $3.0 \, \mu\text{F}$ capacitor with a pd of $12 \, \text{V}$ across it is
$Q = (3.0 \times 10^{-6} \, \text{F})(12 \, \text{V}) = 36 \, \mu\text{C}$

When a second capacitor of $5.0 \, \mu\text{F}$ is connected in parallel, the combined capacitance:
$C = 3.0 \, \mu\text{F} + 5.0 \, \mu\text{F} = 8.0 \, \mu\text{F}$

If the two capacitors are connected in series, their capacitance is:

$$\frac{1}{C} = \frac{1}{3.0 \, \mu\text{F}} + \frac{1}{5.0 \, \mu\text{F}} = \frac{5.0 + 3.0}{15.0 \, \mu\text{F}}$$

$C = 1.9 \, \mu\text{F}$

PRACTICE QUESTIONS

1 Calculate the charge on a 450 pF capacitor when it is connected to a pd of 2.2 kV.

2 a Calculate the combined capacitance of two 9.0 pF capacitors in parallel.

   b The combination is connected in series with a 6.0 pF capacitor. Calculate the combined capacitance.

   c A pd of 12 V is connected across the combination of three capacitors. Calculate:

     i the charge on the 6.0 pF capacitor

     ii the charge on each of the 9.0 pF capacitors.

   d calculate the pd across each of the capacitors.

Energy stored by a capacitor

Work is done to charge up a capacitor. Electrical potential energy is stored.

The work done when a charge is moved through a constant pd, $W = QV$ and when Q is plotted against V the amount of energy transferred is represented by the area under the QV graph.

When a capacitor is charged, the pd across the capacitor increases, as shown in the graph:

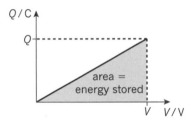

> **REMEMBER:**
> Do not confuse 'Q' with 'C' in equations, this is a common mistake.
> Q is charge in coulombs, C.
> C is capacitance in farads, F.

The energy stored = area under the graph: $W = \frac{1}{2}QV$

Using $Q = CV$ this also gives: $W = \frac{1}{2}CV^2$

WORKED EXAMPLE

The work done when a pd of 9.0 V charges a 250 pF capacitor is:

$W = \frac{1}{2}CV^2$, $C = 250$ pF, $V = 9.0$ V

$W = \frac{1}{2}(250 \times 10^{-12} \text{F})(9.0\text{V})^2 = 1.0 \times 10^{-8}$ J

PRACTICE QUESTIONS

3 Determine how much energy is stored in a 2500 µF capacitor when it is charged by a pd of:

   a 1.5 V   **b** 9.0 V   **c** 12.0 V.

4 A capacitor is connected to a pd of 150 V and stores a 45 µC charge. Calculate the energy stored.

5 Calculate the energy stored on each capacitor in Question 2 above. Check that this is the same as if one capacitor with a capacitance equal to their combined capacitance was used.

10.2 RC circuits 1

Exponential decay

When a capacitor is charged and then discharged through a resistor the rate of discharge is fast at first but keeps reducing as the charge on the capacitor approaches zero, so that the charge never actually reaches zero. The charge, current, and voltage in the circuit all decrease with time as the capacitor discharges.

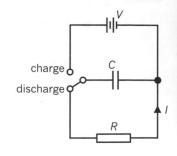

The discharge of a capacitor C through a resistor R is an example of exponential decay (radioactive decay is another example, see Topic 11.1).

Exponential change

The rate of change of something at any time is directly proportional to the amount at that time.

For the charge on a capacitor, the rate of decrease of charge at any time is directly proportional to the charge on the capacitor at that time.

$$\frac{dQ}{dt} = -\frac{Q}{RC}$$

This leads to the equation: $Q = Q_0 e^{-t/RC}$

And also to equations for the voltage and the current:

$$V = V_0 e^{-t/RC} \text{ and } I = I_0 e^{-t/RC}$$

Graphs of V, I, and Q against t are exponential decay curves. This means that, for a graph of Q against t:

- the gradient at any point on the curve is proportional to the value of Q at that point
- equal time intervals result in equal fractional changes in Q.

The same is true for V and I.

Exponential discharge

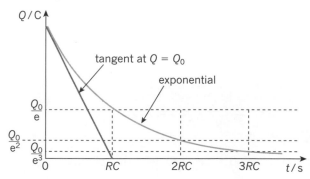

Time constant

The time constant, τ, is the time taken for the charge to fall to $\frac{1}{e}$ of its original value. It depends on the value of C and R: $\tau = RC$

$\frac{1}{e} = e^{-1}$ and can be found using a scientific calculator, it is 0.37 (or 37%) to 2 s.f.

At $t = 0$ the gradient $= -\frac{Q_0}{RC} = -\frac{Q_0}{\tau}$ so the tangent at this point will cut the x-axis at $t = \tau = RC$.

WORKED EXAMPLE

This graph shows the pd across a capacitor as it is discharged through a resistor.

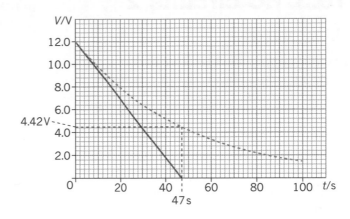

By examining the graph:

The initial pd on the capacitor is 12.0 V

$\frac{1}{e}$ of this value = $0.368 \times 12.0\,V = 4.42\,V$

So the time constant = 47 s

You can also find the time constant from the gradient at $t = 0$ but it is difficult to draw the gradient accurately so, unless it is drawn for you, use the method above.

At $t = 0$ the gradient $= -\dfrac{V_0}{\tau} = -\dfrac{12.0\,V}{47\,s}$ so the time constant = 47 s

PRACTICE QUESTION

1 For the discharge graph in the example, show that the value of V calculated from

$V = \dfrac{V_0}{e^2}$ when $t = 2\tau$ agrees with the value read from the graph.

WORKED EXAMPLE

The pd on a 120 μF capacitor is 36 V. It is discharged through a 250 kΩ resistor. Calculate the pd after 12 s.

$V = ?$, $V_0 = 36\,V$, $C = 120\,\mu F$, $R = 250\,k\Omega$, $t = 8.0\,s$, $V = V_0 e^{-t/RC}$

$V = (36\,V)e^{-(8.0\,s)/(250\,k\Omega)(120\,\mu F)}$

$V = (36\,V)e^{-(0.267)} = 28\,V$

PRACTICE QUESTIONS

2 a Calculate the time constant for an RC circuit with $R = 2.5\,M\Omega$ and $C = 0.40\,\mu F$.

   b Show that the units of the time constant are seconds.

3 A 3200 μF capacitor with a pd across it of 12.0 V is discharged through a 1.0 kΩ resistor. Calculate:

   a the pd across the capacitor after 5.0 s

   b the current in the circuit after 0.50 s (Hint: use $V_0 = I_0 R$ to find I_0.)

   c the charge remaining on the capacitor after 1.0 s. (Hint: use $Q_0 = CV_0$ to find Q_0.)

10.3 RC circuits 2

Natural logs

To calculate the time it takes for the charge on a capacitor to fall to a certain value we rearrange the equation $Q = Q_0 e^{-t/RC}$ to give t.

To do this we take natural logs of both sides of the equation. This is similar to taking logs to base 10 ($\log_{10}$, see Topic 9.3).

Natural logs (logs to base e) are written $\log_e$ or $\ln$

> **REMEMBER:**
> Natural logs
>
> | Example | ln |
> |---|---|
> | $e = e^1$ | $\ln(e) = 1$ |
> | $\dfrac{1}{e} = e^{-1}$ | $\ln(e^{-1}) = -1$ |
> | $1 = e^{1-1} = e^0$ | $\ln(1) = 0$ |
> | 0 | $\ln(0) = -\infty$ |
> | X^A | $\ln(X^A) = \ln(X)$ |
> | $X^A Y^B$ | $\ln(X^A Y^B) = A \ln(X) + B \ln(Y)$ |
> | $\dfrac{X^A}{Y^B} = X^A Y^{-B}$ | $\ln\left(\dfrac{X^A}{Y^B}\right) = A \ln(X) - B \ln(Y)$ |

WORKED EXAMPLE

Calculate the time for the pd across a 120 µF capacitor to fall to 80% of its original value when it is discharged through an 18 kΩ resistor.

You use: $V = V_0 e^{-t/RC}$ where $V = 80\% V_0$

$$\frac{80}{100} V_0 = V_0 e^{-t/(18\,k\Omega)(120\,\mu F)}$$

$$\frac{100}{80} = e^{t/(2.16\,s)}$$

Taking logs of both sides of the equation:

$$\ln(1.25) = \frac{t}{(2.16\,s)}$$

$$t = (2.16\,s)\ln(1.25) = 0.48\,s$$

PRACTICE QUESTIONS

1 Calculate the time for the charge on a 500 µF capacitor to fall to 75% of its initial value when it is discharged through a 12 MΩ resistor.

2 The output from a 1.0 kV power supply must fall to 10.0 V within 6.0 s of being switched off. A 2200 µF capacitor is connected across the terminals inside the power supply. Determine what value resistor is required in series with the capacitor.

Log graphs

The best way to show, by drawing a graph, that the discharge curve is an exponential curve, is to plot a graph of $\ln V$ [or $\ln(I)$ or $\ln(Q)$] against t. This is because it is difficult to say whether a plotted curve is exponential or not, and a log–linear graph will give a straight line.

WORKED EXAMPLE

Show that when this 800 µF capacitor was discharged through a 500 MΩ resistor the pd across the capacitor decreased exponentially with time.

Time t/s	pd V/V	ln(V/V)
0.00	160.00	5.08
0.40	58.90	4.08
0.80	21.70	3.08
1.20	8.00	2.08
1.60	2.93	1.08
2.00	1.08	0.08

Taking logs of both sides of the equation $V = V_0 e^{-t/RC}$

$\ln(V) = \ln(V_0) - \dfrac{t}{RC}$ this is the equation of a straight line $y = mx + c$

where $y = \ln(V)$, $x = t$, gradient $m = -\dfrac{1}{RC}$, and $c = \ln(V_0)$.

$\ln(V) = -\dfrac{t}{RC} + \ln(V_0)$

Plot $\ln(V)$ against t. If they are not given, use your calculator to find the values of $\ln(V)$.

From the graph the time constant

$= -\dfrac{1}{m} = -\dfrac{1}{2.50\,\text{s}}$ gives $RC = 0.40\,\text{s}$

(Note that the line goes through the point (0, 5.08) so this value has been used to calculate the gradient. If, for a line of best fit, the line does not pass through the point, the value used must be the value of the line at $t = 0$.)

Using the values in the question
$RC = (500\,\text{M}\Omega)(800\,\mu\text{F})$
$= (500 \times 10^6\,\Omega)(800 \times 10^{-12}\,\text{F})$
$= 0.40\,\text{s}$.

The straight line shows that the discharge is exponential with time constant $RC = 0.40\,\text{s}$.

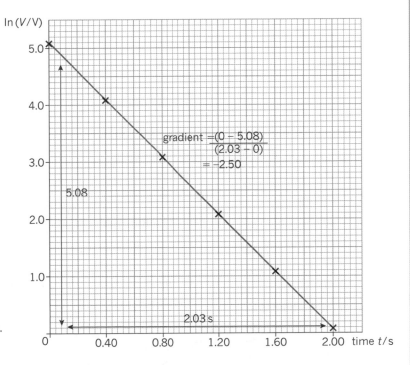

PRACTICE QUESTION

3 Show that this data for the discharge of a capacitor is exponential and find the time constant for the circuit.

Time t/s	0.0	20.0	40.0	60.0	80.0	100.0
pd V/V	4.06	3.02	2.25	1.68	1.25	0.93

11 RADIOACTIVITY

11.1 Radioactive decay

Exponential decay

All nuclei are made up of a number of protons and neutrons (hydrogen has one proton). Some of these combinations are stable, whilst others are unstable and may spontaneously disintegrate. These unstable nuclei are radioactive and the disintegration is radioactive decay. All nuclei with more than 83 protons are unstable.

Radioactive decay is random and unpredictable — it is impossible to tell when a nucleus may decay — but for each type of nucleus (each isotope) the probability of decay is constant. This means that, for a large sample of those nuclei, we can say how many, on average, will decay in a certain time.

The number of nuclei that decay in a certain time will depend on how many are present. This is an example of an exponential decay (see also Topic 10.2).

The rate of decrease of something at any time is directly proportional to the amount at that time.

For a number of radioactive nuclei of one isotope, the rate of decrease at any time is directly proportional to the number of undecayed nuclei, N, at that time.

$$\frac{dN}{dt} = -\lambda N$$

This leads to the equation: $N = N_0 e^{-\lambda t}$ **(1)**

λ is the radioactive decay constant.

λ is measured in units of s^{-1}, when t is in seconds.

The value of λ is different for different isotopes. If it is large, the isotope is one that decays more quickly.

λ from a graph

The radioactive decay constant can be found from the gradient of a graph of $\ln(N)$ against t:

Taking natural logs of equation **(1)** (see Topic 10.3)

$\ln(N) = \ln(N_0) - \lambda t$

This is the equation of a straight line where $y = \ln(N)$, $x = t$, gradient $m = -\lambda$, and $c = \ln(N_0)$.

Graph of ln(N) against t

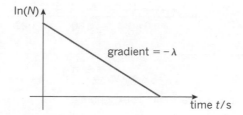

WORKED EXAMPLE

A sample contains 1.2×10^{12} polonium-210 atoms. Polonium-210 has a radioactive decay constant of 5.0×10^{-3} days. Calculate the number of atoms left after 4 weeks.

4 weeks = 56 days and because we are calculating λt we can leave λ in day^{-1} and t in days.

$\lambda t = (5.0 \times 10^{-3} day^{-1})(56\ days) = 0.28$

$N = (1.2 \times 10^{12})e^{-0.28} = 9.1 \times 10^{11}$

PRACTICE QUESTIONS

1 The radioactive decay constant for carbon-14 is 1.2×10^{-4} year^{-1}. Calculate how many of a sample of 15 000 atoms will be left after 22 920 years.

2 A radioactive source has a decay constant λ of 2.0×10^{-4} s^{-1}. After exactly 5 minutes there are 25 000 nuclei left. Calculate how many nuclei were present at the start.

Half-life

The half-life, $T_{\frac{1}{2}}$ or $t_{\frac{1}{2}}$, of a radioactive isotope is the average time taken for half of the radioactive nuclei to decay.

From whatever point you start on the decay curve, one half-life later the number of radioactive nuclei will have halved.

In equation **(1)** on the previous page, $T_{\frac{1}{2}}$ is the time for the number of nuclei N_0 to fall to $\frac{N_0}{2}$.

Substituting and taking natural logs:

$\ln\frac{1}{2} = -\lambda T_{\frac{1}{2}}$

$\ln(2) = \lambda T_{\frac{1}{2}}$ which may be given as $T_{\frac{1}{2}} = \frac{\ln(2)}{\lambda}$ or $\lambda = \frac{\ln(2)}{T_{\frac{1}{2}}}$

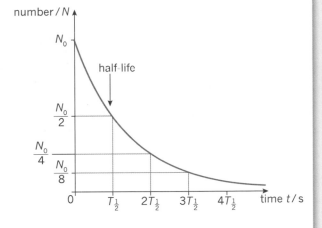

WORKED EXAMPLE

The half-life of radon-226 is 1.6×10^3 years. If a sample contains 25×10^6 nuclei, calculate the number left after 5.0×10^3 years.

$\lambda = \frac{\ln(2)}{(1.6 \times 10^3\,y)} = 4.33 \times 10^{-4}\,y^{-1}$

$\lambda t = (4.33 \times 10^{-4}\,y^{-1})(5.0 \times 10^3\,y) = 2.17$

$N = (25 \times 10^6)e^{-2.17} = 2.9 \times 10^6$

PRACTICE QUESTIONS

3 The decay constant for uranium-238 is $1.55 \times 10^{-10}\,y^{-1}$. Calculate the half-life.

4 The half-life for technetium-99m is 6.01 hours. Calculate the radioactive decay constant in s^{-1}.

5 A sample contains 4.0 g of gold-199. How much will be left after 10.0 days? (The half-life of gold-199 = 3.15 days.)

6 Sketch a decay curve showing the percentage of iodine-131 that remains in a sample that starts with 100% iodine-131 at time $t = 0$. (Iodine-131 has a half-life of 8.0 days.)

STRETCH YOURSELF!

The ratio of the number of uranium-238 atoms to the number of lead-206 atoms in a rock is 0.42. The rock originally contained no lead-206. Determine its age. (The decay constant for uranium-238 decay to Pb-206 is $1.55 \times 10^{-10}\,y$.)

11.2 Activity

The activity, A, of a radioactive source is the number of disintegrations per unit time.

$$A = \frac{dN}{dt}$$

Using $\frac{dN}{dt} = -\lambda N$ this gives $A = -\lambda N$ (This is the equation used to define λ.)

Substituting for N:

$$A = -\lambda N_0 e^{-\lambda t}$$

The activity at $t = 0$ is $A_0 = -\lambda N_0$ so:

$$A = A_0 e^{-\lambda t}$$

The unit of activity is the bequerel, Bq. $1\,Bq = 1$ disintegration s^{-1}.

The activity also gives us another way to define half-life:

The half-life is the average time for the activity of a sample of a radioactive isotope to halve.

Radioactive decay and half-life

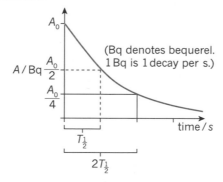

WORKED EXAMPLE

A radioactive source with half-life 28 years emitted 1.0×10^4 beta particles per second in 1993. Calculate the activity in 2014.

$A_0 = 1.0 \times 10^4\,Bq \quad T_{\frac{1}{2}} = 28\,y \quad t = (2014-1993) = 21\,y$

$\lambda = \dfrac{\ln(2)}{T_{\frac{1}{2}}}$ (see Topic 11.1)

$\lambda = \dfrac{\ln(2)}{(28\,y)} = 0.0248\,y^{-1}$

$\lambda t = (0.0248\,y^{-1})(21\,y) = 0.521$

$A = (1.0 \times 10^4\,Bq)e^{-(0.521)} = 5900\,Bq$

PRACTICE QUESTIONS

1   The half-life of radium-226 is 1.6×10^3 years. A sample contains 3.2×10^{16} nuclei. Calculate the activity.

2   A sample of technetium-99m (half-life = 6.01 h) has an activity of $1.1 \times 10^4\,Bq$ when it is prepared. Calculate the activity when it is used 2 hours later.

3   Sodium-24 has a half-life of 15 hours. A sample has an initial activity of 75 Bq. Calculate the activity after 8 hours.

4 A 50.0 g sample is taken from a skeleton and has an activity, due to the carbon-14 content, of 1.2×10^3 Bq. The activity for a living organism is 9.0×10^2 Bq per g. Carbon-14 has a half-life of 5730 y. Calculate the age of the skeleton.

STRETCH YOURSELF!

Source X has a half-life of 100 years and source Y has a half-life of 10 years. Their activities are initially the same. Compare their activities:

a 6 months later

b 6 years later.

WORKED EXAMPLE

Use the following data to plot a straight-line graph to determine the radioactive decay constant and the half-life for the radioactive isotope.

This can be done by plotting ln(A) against t, (in the same way as ln(N) against t, see Topic 11.1).

Time t/s	Activity A/ ×10⁴ Bq	Ln(A/Bq)
0	3.63	10.5
300	2.97	10.3
600	1.99	9.9
900	1.34	9.5
1200	0.99	9.2
1500	0.73	8.9
1800	0.54	8.6
2100	0.40	8.3
2400	0.27	7.9

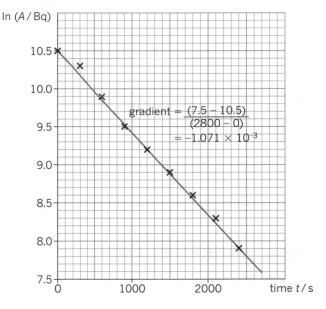

The radioactive decay constant $\lambda = -$gradient

$$= 1.07 \times 10^{-3}\,\text{s}^{-1}$$

The half-life $= \dfrac{\ln(2)}{\lambda} = \dfrac{\ln(2)}{(1.07 \times 10^{-3}\,\text{s}^{-1})} = 650\,\text{s}$

PRACTICE QUESTION

5 Use the following data to plot a straight-line graph to determine the radioactive decay constant and the half-life for the radioactive isotopes.

Time t/hours	0	0.5	1.0	1.5	2.0
Activity A/ ×10⁷ Bq	8.18	7.25	6.52	5.71	5.08

12 NUCLEAR PHYSICS

12.1 Nuclear reactions

Nuclear equations

A nuclide is any particular version of a nucleus. When a nuclide decays some quantities are conserved. The total energy is conserved (see next section) and momentum is conserved when no external forces act on a system. Electric charge is also conserved.

In equations, a nuclide is represented as:

$_Z^A X$ ← chemical symbol of element

↗ nucleon number (number of protons + neutrons)

↖ proton number (number of protons)

The proton number, Z, is a measure of the positive charge. For example, an electron is given a proton number of -1 because it has an equal and opposite charge to the proton. A neutron has a proton number of 0.

The nucleon number A (also called the mass number) is 1 for protons and neutrons and 0 for particles like the electron.

In nuclear reactions, A and Z are conserved.

WORKED EXAMPLE
Alpha decay

An alpha particle is a fast-moving helium nucleus with $A = 4$ and $Z = 2$.
When a radium-226 nucleus emits an alpha particle, the decay product will have $A = 226 - 4 = 222$ and $Z = 88 - 2 = 86$. The element with 86 protons is radon:

$$_{88}^{226}\text{Ra} \rightarrow\ _{86}^{222}\text{Rn} +\ _2^4\text{He}$$

WORKED EXAMPLE
Beta decay

A beta particle is a fast-moving electron emitted by the nucleus with $A = 0$ and $Z = -1$. When a boron-12 nucleus emits a beta particle, the decay product will have $A = 12 - 0 = 12$ and $Z = 5 - (-1) = 6$. The element with 6 protons is carbon. An anti-neutrino (see Topic 13.1) is also emitted and this has $A = 0$ and $Z = 0$:

$$_5^{12}\text{Bo} \rightarrow\ _6^{12}\text{C} +\ _{-1}^0\text{e} +\ _0^0\bar{\nu}$$

PRACTICE QUESTIONS

1 Write equations for the following reactions:
   a Plutonium-240 ($Z = 94$) decays to uranium by alpha emission.
   b Carbon-14 ($Z = 6$) decays to nitrogen by beta emission.

2 When an alpha particle strikes a nitrogen-14 ($Z = 7$) nucleus they combine and then decay to an oxygen-17 nucleus ($Z = 8$) and another particle.
   Write a nuclear equation and say what the other particle is.

$E = mc^2$

From the theory of relativity Einstein concluded that energy, E, and mass, m, were equivalent. This is expressed in the equation $E = mc^2$ or sometimes $\Delta E = \Delta mc^2$, where c is the speed of light in a vacuum $= 3.00 \times 10^8 \, \mathrm{m\,s^{-1}}$.

In nuclear reactions, you must take into account that the total of the energy + mass is conserved, but that energy may be transferred to mass, or mass to energy, according to $\Delta E = \Delta mc^2$. When a particle is not moving the mass is referred to as its rest mass m_0. Its mass increases with speed, but this is insignificant until the speed approaches the speed of light.

When a nuclear reaction takes place, the mass of the products may be different to the mass of the original nuclides and this mass defect is the mass transferred as energy.

Mass defect $= \sum$ mass of original nuclides $- \sum$ mass of products
(The symbol '$\sum$' means the sum of all.)

The atomic mass unit

The mass of the proton is $1.673 \times 10^{-27} \, \mathrm{kg}$. To avoid working with such small numbers the atomic mass unit u is used. It is the mass of $\frac{1}{12}$ of one atom of the carbon–12 isotope: $1 \, \mathrm{u} = 1.661 \times 10^{-27} \, \mathrm{kg}$.

WORKED EXAMPLE

Calculate the mass defect in the decay of the radium nucleus on the previous page and the energy released.

Masses: radium-226 nucleus $= 226.0254 \, \mathrm{u}$, radon-222 nucleus $= 222.0175 \, \mathrm{u}$, helium-4 nucleus $= 4.0026 \, \mathrm{u}$

$226.0254 - (222.0175 + 4.0026) = 0.0053 \, \mathrm{u}$

This is positive, so energy is transferred to the kinetic energy of the products.

This is $0.0053 \times 1.661 \times 10^{-27} \, \mathrm{kg} = 8.80 \times 10^{-30} \, \mathrm{kg}$

$\Delta E = \Delta mc^2 = (8.80 \times 10^{-30} \, \mathrm{kg})(3.00 \times 10^8)^2$

$\Delta E = 7.92 \times 10^{-13} \, \mathrm{J}$

This is a very small energy and the unit eV is usually used (see Topic 6.1). $1 \, \mathrm{eV} = 1.6 \times 10^{-19} \, \mathrm{J}$. When electrons are accelerated, typical energies are keV.
For nuclear reactions typical energies are MeV. $1 \, \mathrm{MeV} = 10^6 \, \mathrm{eV}$

$$\Delta E = \frac{7.92 \times 10^{-13} \, \mathrm{J}}{10^6 \times 1.6 \times 10^{-19} \, \mathrm{J\,MeV^{-1}}} = 4.95 \, \mathrm{MeV}$$

PRACTICE QUESTIONS

3 Calculate, in atomic mass units, the mass of:
   a the proton
   b the neutron
   c the electron.   (See formulae sheet at the end of the book.)
      Give your answers to 4 s.f.

4 Polonium-210 ($Z = 84$) decays to lead by alpha-particle emission.
   a Give the nuclear equation for the decay.
   b Calculate the mass defect in u.
   c Calculate the energy released in MeV.
      Mass: Po-210 $= 209.937 \, \mathrm{u}$ Pb-206 $= 205.929 \, \mathrm{u}$ He-4 $= 4.002 \, \mathrm{u}$

> **REMEMBER:** When calculating the mass defect, keep all the decimal places when adding and subtracting. The mass defect is very small and often depends on $0.0001 \, \mathrm{u}$.

12.2 Fission and fusion

Binding energy

The binding energy per nucleon is the energy needed to separate a nucleus into separate protons and neutrons. The stability of the nucleus depends on the binding energy per nucleon. Although very large nuclei have more binding energy, their binding energy per nucleon is less. The most stable nucleus is the iron–56 nucleus.

> **REMEMBER:** A nucleus has a lower mass than its separate protons and neutrons. It is more stable. It must be given energy to separate the nucleons.
>
> The phrase 'binding energy' often causes confusion because it sounds like something added to the nucleus to hold it together – some people say it would be easier to think of it as 'unbinding energy'.

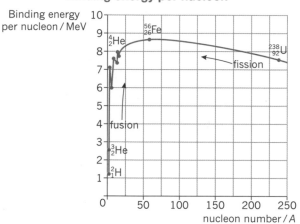

Binding energy per nucleon

The binding energy is found by calculating the mass defect between the total mass of the separate protons and neutrons and the mass of the nucleus.

WORKED EXAMPLE

The mass of the lithium-7 ($Z = 3$) nucleus is 7.016 u. Find its binding energy.

$$\text{Mass of proton (from formulae sheet)} = \frac{1.673 \times 10^{-27}\,\text{kg}}{1.661 \times 10^{-27}\,\text{kg u}^{-1}} = 1.007\,\text{u}$$

$$\text{Mass of neutron} = \frac{1.675 \times 10^{-27}\,\text{kg}}{1.661 \times 10^{-27}\,\text{kg u}^{-1}} = 1.008\,\text{u}$$

3 protons + 4 neutrons = $3(1.007\,\text{u}) + 4(1.008\,\text{u}) = 7.053\,\text{u}$

Mass defect = $7.053 - 7.016 = 0.037\,\text{u}$

Binding energy = $(0.037\,\text{u})(1.661 \times 10^{-27}\,\text{kg u}^{-1})(3.00 \times 10^8\,\text{m s}^{-1})^2\,\text{J} = 5.53 \times 10^{-12}\,\text{J}$

$$\text{Binding energy in MeV} = \frac{5.53 \times 10^{-12}\,\text{J}}{1 \times 10^6 \times 1.60 \times 10^{-19}\,\text{J MeV}^{-1}} = 35\,\text{MeV}$$

Alternatively, changing mass in u to an equivalent energy in MeV can be done by using the conversion factor: 1 u = 931.5 MeV.

PRACTICE QUESTIONS

1 a Use the masses for the proton and neutron to calculate a mass for the helium-4 nucleus.
   b The actual mass of the helium-4 nucleus is 4.00 u (3 s.f.). Calculate the mass defect in u.
   c The difference in mass is the binding energy. Calculate this in MeV.

2 For the iron-56 ($Z = 26$) nucleus, calculate:
   a the mass defect   b the binding energy
   c the binding energy per nucleon. (Mass of iron-56 = 55.935 u.)

Fission

As shown on the graph of binding energy per nucleon, when nuclei more massive than the iron-56 nucleus are split into two nuclei of roughly similar size, the products will have less binding energy per nucleon and there will be a release of energy. This is the principle behind nuclear reactors and nuclear fission bombs.

WORKED EXAMPLE

$$^{235}_{92}U + ^{1}_{0}n \rightarrow ^{144}_{56}Ba + ^{90}_{36}Kr + 2^{1}_{0}n$$

mass uranium-235 = 235.0439 u

mass barium-144 = 143.9230 u

mass krypton-90 = 89.9195 u

mass neutron = 1.0087 u

Initial mass = 235.0439 u + 1.0087 u

Final mass = 143.9230 u + 89.9195 u + 2 × (1.0087 u)

Mass defect = 235.0439 u − (143.9230 u + 89.9195 u + 1.0087 u)

Mass defect = 235.0439 u − (234.8512) = 0.1927 u

Energy released = 0.1927 × 931.5 MeV

Energy = 180 MeV

Fusion

Also shown by the graph of binding energy per nucleon, when nuclei much less massive than iron-56 are joined by nuclear fusion the product is more stable and there is a release of energy. This is the process that releases energy in stars and nuclear fusion bombs, and is the principle behind experimental nuclear fusion reactors.

WORKED EXAMPLE

$$^{2}_{1}H + ^{3}_{1}H \rightarrow ^{4}_{2}He + ^{1}_{0}n$$

mass hydrogen-2 = 2.0141 u

mass hydrogen-3 = 3.0160 u

mass helium-4 = 4.0026 u

mass neutron = 1.0087 u

Initial mass = 2.0141 u + 3.0160 u = 5.0301 u

Final mass = 4.0026 u + 1.0087 u = 5.0113 u

Mass defect = 5.0301 u − 5.0113 u = 0.0188 u

Energy released = 0.0188 × 931.5 MeV = 17.5 MeV

PRACTICE QUESTION

3 Calculate the energy released in the following reactions:

a $^{235}_{92}U + ^{1}_{0}n \rightarrow ^{148}_{57}La + ^{85}_{35}Br + 3^{1}_{0}n$

mass uranium-235 = 235.0439 u

mass lanthanum-148 = 147.9322 u

mass bromine-85 = 84.9156 u

mass neutron = 1.0087 u

b $^{14}_{7}N + ^{1}_{1}H \rightarrow ^{15}_{8}O + ^{0}_{0}\gamma$

mass nitrogen-14 = 14.0031 u

mass hydrogen-1 = 1.0078 u

mass oxygen-15 = 15.0031 u

(Gamma rays have rest mass = 0 and charge = 0.)

13.1 Fundamental particles 1

The standard model

In the standard model of particle physics there are 12 fundamental particles from which all matter is made and 12 corresponding antiparticles from which all antimatter is made. Fundamental particles are particles that cannot be divided into smaller particles.

There are six leptons and six quarks (and six anti-leptons and six anti-quarks).

We can represent particle interactions with equations. There are a number of conservation rules that describe which interactions are allowed.

Some describe conservation of:
- mass/energy
- momentum
- electric charge
- lepton number (see below)
- baryon number (see below)

Leptons

Leptons have small rest mass. (Neutrinos were thought until recently to have zero rest mass.)

Lepton number is conserved.

	Leptons		Anti-leptons	
Lepton number	1	1	−1	−1
Charge/e	−1	0	+1	0
1st generation	electron e^-	electron-neutrino v_e	positron e^+ or $\bar{e}$	electron-anti-neutrino $\bar{v}_e$
2nd generation	muon μ^-	muon-neutrino v_μ	anti-muon μ^+ or $\bar{\mu}$	muon-anti-neutrino $\bar{v}_\mu$
3rd generation	tau or tauon τ^-	tau-neutrino v_τ	anti-tau τ^+ or $\bar{\tau}$	tau-anti-neutrino $\bar{v}_\tau$

WORKED EXAMPLE

A neutron is unstable and can decay to a proton and an electron. This is β^- decay. Conservation of lepton number requires an anti-lepton to be formed along with the electron (a lepton). Conservation of charge means it must be an anti-neutrino, and as it is formed with the electron it will be an electron-anti-neutrino:

$$n \rightarrow p + e^- + \bar{v}_e$$

Charge:	(0) → (+1) + (−1) + (0)	✓ conserved
Lepton no.:	(0) → (0) + (+1) + (−1)	✓ conserved

> **REMEMBER:** Check your specification to see which particles you need to know about.
>
> Check whether information is on your formulae sheet.

PRACTICE QUESTIONS

1 Give a particle equation for β^+ decay, when a proton decays to give a neutron and a positron. (Note: this can only happen when energy is given to the proton, for example when it is inside a nucleus.)

2 A muon is unstable and can decay to an electron, an electron-anti-neutrino, and another particle. Give a particle equation and use it to find out the other particle.

Quarks

Quarks have never been found to exist on their own. They form particles called hadrons. There are two types of hadron:

- Baryons are made of three quarks (and anti-baryons of three anti-quarks).
- Mesons are made of a quark and an anti-quark.

There is something about the strange quark that is conserved in reactions involving the strong force and the electromagnetic force; this has been named strangeness. The strange quark has a strangeness of -1 and the anti-strange quark a strangeness of $+1$. All other quarks have a strangeness of 0.

	Quarks		Anti-quarks	
Charge/e	$+\frac{2}{3}$	$-\frac{1}{3}$	$-\frac{2}{3}$	$+\frac{1}{3}$
Baryon number	$+\frac{1}{3}$	$+\frac{1}{3}$	$-\frac{1}{3}$	$-\frac{1}{3}$
1st generation	up u	down d	anti-up $\bar{u}$	anti-down $\bar{d}$
2nd generation	charmed c	strange s	anti-charmed $\bar{c}$	anti-strange $\bar{s}$
3rd generation	top t	bottom b	anti-top $\bar{t}$	anti-bottom $\bar{b}$

REMEMBER:
Even the strangeness number is strange; it is -1 for the strange quark and $+1$ for the anti-strange quark (baryon and lepton numbers are $+$ for matter and $-$ for antimatter).

The neutron is made of an up quark and two down quarks: n = udd.

The proton is made of two up quarks and a down quark: p = uud.

WORKED EXAMPLE

Show that the combination of one up and two down quarks has the correct charge, lepton number, and baryon number to be a neutron.

A neutron has a charge = 0, lepton number = 0, and baryon number = +1.

	u	d	d
Charge:	$(+\frac{2}{3})$ +	$(-\frac{1}{3})$ +	$(-\frac{1}{3})$ = 0
Lepton no.:	(0) +	(0) +	(0) = 0
Baryon no.:	$(+\frac{1}{3})$ +	$(+\frac{1}{3})$ +	$(+\frac{1}{3})$ = +1

PRACTICE QUESTIONS

3 Show that there are three more combinations of up and down and anti-up and anti-down quarks. Also show that these have the correct charge, lepton number, and baryon number to be a proton, an anti-proton, and an anti-neutron.

4 Draw a table to show the charge, strangeness, lepton number, and baryon number of these particles:

   a an omega (Ω) made of three strange quarks

   b a lambda (λ) made of an up, down, and strange quark

   c each of the three sigma (Σ) particles uus, uds, and dds.

13.2 Fundamental particles 2

Mesons

Mesons are made of a quark and an anti-quark.

They will all have baryon numbers $(+\frac{1}{3}) + (-\frac{1}{3}) = 0$

WORKED EXAMPLE

Mesons called pions, symbol π, are made of combinations of up and down quarks and anti-quarks. List the types of pion. Work out the charge on each one. Mesons are a quark and an anti-quark. Possible combinations are: $\bar{u}u$ $\bar{d}d$ and $u\bar{d}$ $d\bar{u}$.

$\bar{u}u$: charge $= (+\frac{2}{3}) + (-\frac{2}{3}) = 0\ \pi^0$

$\bar{d}d$: charge $= (-\frac{1}{3}) + (+\frac{1}{3}) = 0\ \pi^0$

$u\bar{d}$: charge $= (+\frac{2}{3}) + (+\frac{1}{3}) = +1\ \pi^+$

$d\bar{u}$: charge $= (-\frac{1}{3}) + (-\frac{2}{3}) = -1\ \pi^-$

PRACTICE QUESTIONS

1 Mesons called kaons, symbol K, contain the strange quark or anti-quark, together with up or down quarks or anti-quarks.

   a List the types of kaon using symbols.

   b Calculate the charge on each one.

   c Give the strangeness number for each one.

2 A π^0 decays to an electron, a positron, and two gamma photons. Write a particle equation for the decay and show that charge, lepton number, and baryon number are conserved.

3 The K^- particle ($u\bar{s}$) decays to a pion, an electron, and an electron-anti-neutrino.

   a Write a particle equation for the decay.

   b Determine the charge on the pion.

   c Show that strangeness is not conserved in this decay.

Matter and antimatter

When a particle meets its antiparticle the quantities such as charge, strangeness, and baryon number cancel and it is annihilated. There is no negative mass, so the mass becomes the energy of two photons to conserve mass/energy. There are two photons so that momentum is conserved.

Energy can also be converted into mass.

WORKED EXAMPLE
Pair production

A high-energy X-ray or gamma ray photon passes close to a nucleus and interacts with the nucleus. Its energy is absorbed and an electron and positron are emitted in a process called pair production. Calculate the minimum energy of the photon for which this process can occur.

The minimum energy will be that of a stationary electron e^- and a stationary positron e^+. This is the energy equivalent to the rest mass of the e^- and the rest mass of the e^+.

The rest mass of the e^- is equivalent to $\Delta E = \Delta mc^2$ where $\Delta m = 9.11 \times 10^{-31}$ kg and $c = 3.00 \times 10^8$ m s^{-1}

$\Delta E = (9.11 \times 10^{-31}$ kg$)(3.00 \times 10^8$ m s$^{-1})^2 = 8.2 \times 10^{-14}$ J

$\Delta E = \dfrac{(8.2 \times 10^{-14} \text{ J})}{10^6 \times (1.60 \times 10^{-19} \text{ J eV}^{-1})} = 0.51$ MeV

The rest mass of the e^+ is the same as that of the e^- so the minimum energy of the photon is:

$\Delta E = 2 \times (0.51 \text{ MeV}) = 1.02$ MeV

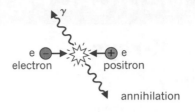

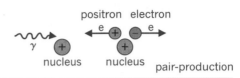

Units MeV/c^2

Particle physicists work with energies of MeV or higher (GeV or TeV).

From the above example the rest mass of an electron or positron is equivalent to 0.51 MeV of energy. It would be incorrect to say that the rest mass of a proton or an electron is 0.51 MeV because the electron-volt is a unit of energy. However, using the fact that $\Delta E = \Delta mc^2$ you can say that the rest mass of the electron or positron is 0.51 MeV $/c^2$.

This unit is widely used in particle physics, just as the eV is widely used by physicists working with electron beams. (An electron beam produced by accelerating electrons using a pd of 1.0 kV produces electrons of energy 1.0 keV; see Topic 6.1.)

WORKED EXAMPLE

Show that 1 MeV/c^2 is equivalent to a mass of about 1.8×10^{-30} kg.

$\Delta m = 1 \text{ MeV}/c^2 = \dfrac{1 \times 10^6 \times (1.60 \times 10^{-19} \text{ J eV}^{-1})}{(3.00 \times 10^8 \text{ m s}^{-1})^2} = 1.78 \times 10^{-30}$ kg

Note that in a 'show that' question you should give one extra significant figure in your answer.

PRACTICE QUESTIONS

4 A gamma ray photon creates an electron–positron pair each with kinetic energy of 0.25 MeV. Calculate the energy of the photon.

5 Calculate the rest mass in MeV/c^2 of:
   a the proton   **b** the neutron.

6 Give the equivalent mass of 1 u in MeV/c^2.

7 Determine the total kinetic energy of an electron–positron pair if they are created from a 2.5 MeV photon.

8 An electron and a positron both moving in opposite directions with energy of 5.0 MeV collide. They annihilate and produce two gamma ray photons moving in opposite directions. Calculate the wavelength of each photon.

STRETCH YOURSELF!

Determine the minimum photon energy required to produce a $\mu^+ \mu^-$ pair.
(μ rest mass $= 1.9 \times 10^{-28}$ kg.)

14 MEDICAL PHYSICS

14.1 X-rays

Intensity

When any radiation source, for example, of light, X–rays, or gamma rays, radiates power, P, uniformly in all directions, it will be evenly spread over the surface of a sphere.

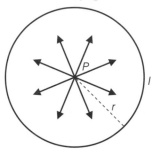

The surface area of a sphere is $4\pi r^2$, so the intensity I of the radiation at a distance r from the source is given by:

$$\text{intensity} = \frac{\text{power radiated}}{\text{area of surface}} \qquad I = \frac{P}{4\pi r^2}$$

The intensity is measured in watts per square metre: W m^{-2}.

If the intensity, I, of a beam of radiation is collimated (a parallel beam) and is falling on a surface of cross-sectional area A, the power, P, incident on the surface is:

$$P = IA$$

WORKED EXAMPLE

A collimated X-ray beam of intensity $5.2 \times 10^8\,\text{W m}^{-2}$ is incident on an area of $1.0\,\text{cm}^2$. Calculate the energy per second on the surface.

Energy per second $= P = IA = (5.0 \times 10^8\,\text{W m}^{-2})(1.0 \times 10^{-2} \times 10^{-2}\,\text{m}^2)$

Energy per second $= 52\,\text{kW} = 52\,\text{kJ s}^{-1}$

PRACTICE QUESTIONS

1 A collimated X-ray beam of intensity $4.5 \times 10^8\,\text{W m}^{-2}$ is incident on a spot of radius $2.5\,\text{cm}$. Calculate the power incident on the surface.

2 An X-ray beam of intensity $6.0 \times 10^8\,\text{W m}^{-2}$ forms a collimated beam of circular cross section. Calculate the radius of the area needed for the energy absorbed by the surface in $2.0\,\text{s}$ to be $4.6\,\text{MJ}$.

3 The intensity of X-rays from the Sun reaching the upper atmosphere of the Earth varies according to the Sun's activity. When the intensity is $6.0 \times 10^{-7}\,\text{W m}^{-2}$, calculate the total power of the X-rays emitted from the Sun.
   (Sun–Earth distance $= 1.5 \times 10^{11}\,\text{m}$)

Attenuation

When a collimated X-ray beam passes through a material the energy is absorbed and the intensity is reduced. The amount absorbed will depend on the intensity of the beam, so this is an example of an exponential decrease.

In this case, the intensity I decreases exponentially with the distance x travelled through the material. (For examples of exponential changes with time and natural logs see Topics 10.2 and 11.1.)

$$\frac{dI}{dx} = -\mu I \text{ and } I = I_0 e^{-\mu x}$$

Taking natural logs $\ln(I) = \ln(I_0) - \mu x$

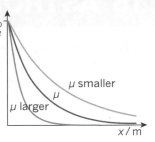

μ is the absorption coefficient measured in m^{-1}. If μ is large the radiation is strongly attenuated, it will not travel far in the material.

The half value thickness (HVT) $x_{\frac{1}{2}}$ is the thickness of the material that reduces the intensity to half of its original value $\frac{I_0}{2}$.

$$\ln(\tfrac{1}{2}) = -\mu x_{\frac{1}{2}}$$

$$x_{\frac{1}{2}} = \frac{\ln 2}{\mu}$$

This is similar to the half–life for radioactive decay (see Topic 11.1).

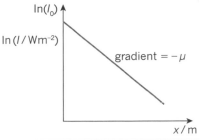

WORKED EXAMPLE

An X-ray beam has an intensity of $6.0 \times 10^8\,W\,m^{-2}$. Calculate the intensity after travelling through a 5.0 mm thick aluminium plate.

(The absorption coefficient for aluminium is $\mu_{Al} = 1.5\,mm^{-1}$.)

$I_0 = 6.0 \times 10^8\,W\,m^{-2}$ $x = 5.0\,mm$

$\mu x = (1.5\,mm^{-1})(5.0\,mm) = 7.5$

$I = (6.0 \times 10^8\,W\,m^{-2})e^{-7.5}$

$I = 3.3 \times 10^5\,W\,m^{-2}$

WORKED EXAMPLE

The intensity of a collimated X-ray beam is reduced to 75% of its initial value when it passes through 1.6 cm of muscle. Calculate the half-value thickness for muscle.

$I = 75\%\,I_0$ $x = 1.6\,cm$

$0.75\,I_0 = I_0 e^{-\mu(1.6\,cm)}$

$\ln(0.75) = -\mu(1.6\,cm)$

$\mu = \dfrac{-\ln(0.75)}{(1.6\,cm)} = 0.180\,cm^{-1}$

$x_{\frac{1}{2}} = \dfrac{\ln(2)}{(0.180\,cm^{-1})} = 3.9\,cm^{-1}$

PRACTICE QUESTIONS

4 A collimated X-ray beam has an intensity of $4.8 \times 10^8\,W\,m^{-2}$. Calculate the intensity after travelling through a 3.0 mm thick aluminium plate. (The absorption coefficient for aluminium is $\mu_{Al} = 1.5\,mm^{-1}$.)

5 The intensity of a collimated beam of X-rays is reduced to 12% of its initial value when it passes through 4.0 mm of soft tissue. Calculate the thickness that reduced the intensity to 20% of its initial value.

15.1 Red shift

The Doppler effect

The Doppler effect was first observed with sound waves and occurs with all waves, including all electromagnetic radiation.

When there is relative motion between a source of waves and an observer, the wavelength and frequency detected by the observer changes.

If the source and observer are getting closer together the wavelength decreases and the frequency increases.

If the source and observer are getting further apart the wavelength increases and the frequency decreases.

The faster the relative velocity is, the greater the change in frequency and wavelength.

Note that the source or the observer may be stationary or both may be moving. It is the relative velocity between them that determines the effect.

This diagram shows the frame of reference where the observer is stationary. Source A is stationary and observer 1 detects waves with wavelength λ coming from source A. Source B is moving away from observer 1, who detects a longer wavelength. Source B is moving towards observer 2, who detects a shorter wavelength.

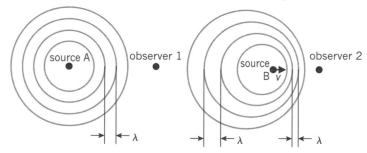

Using c for the speed of the waves and v for the relative speed between the observer and the source:

Doppler shift, z, in wavelength $\lambda = \triangle\lambda$ and Doppler shift in frequency $f = \triangle f$

When $v \gg c$

$$z = \frac{\triangle f}{f} = -\frac{\triangle\lambda}{\lambda} = \frac{v}{c}$$

✓ WORKED EXAMPLE

The spectrum of light from a star is shifted so that a spectral line at 656.2 nm appears to be at 655.8 nm. Calculate the value of z and the velocity of the star. State whether it is moving towards or away from the Earth.

$\lambda = 656.2\,\text{nm}$ $\triangle\lambda = (656.2\,\text{nm} - 655.8\,\text{nm}) = 0.4\,\text{nm}$

The shift is to shorter λ so the star is moving towards the Earth, and z is negative.

$$z = -\frac{\triangle\lambda}{\lambda} = -\frac{(0.4\,\text{nm})}{(656.2\,\text{nm})} = -6.10 \times 10^{-4} = -6.1 \times 10^{-4}\ (2\ \text{s.f.})$$

$v = zc = -(6.10 \times 10^{-4})(3.00 \times 10^{8}\,\text{m s}^{-1}) = -1.8 \times 10^{5}\,\text{m s}^{-1}$

$(1.8 \times 10^{5}\,\text{m s}^{-1}$ towards the Earth)

PRACTICE QUESTION

1 The Andromeda galaxy has $z = -0.001$.
a Calculate its velocity.
b Determine the wavelength of the sodium spectral line at 589.59 nm in light from the Andromeda galaxy observed on Earth.

Red and blue shifts

A shift in the frequency or wavelength is referred to as a red shift if it is to longer wavelengths (lower frequencies) and a blue shift if it is to shorter wavelengths (higher frequencies). Note that a microwave wavelength with a red shift would be shifted towards the longer radio wavelengths – not towards red light. Similarly, an ultraviolet wavelength with a blue shift would be shifted towards X-rays, not towards blue light.

WORKED EXAMPLE

In light from a galaxy, an ultraviolet spectral line at 253.7 nm is shifted to 254.2 nm. Explain whether the galaxy is moving towards or away from Earth. Calculate a value for z. Identify whether this is a blue or red shift.

The shift is to a longer wavelength so the galaxy is moving away from Earth and this is a red shift.

$$z = -\frac{\triangle\lambda}{\lambda} = -\frac{(254.2\,\text{nm} - 253.7\,\text{nm})}{(253.7\,\text{nm})} = 2.0 \times 10^{-3}$$

PRACTICE QUESTIONS

2 The radio signals from a distant galaxy are red shifted by 12% from their values on Earth. Calculate the speed of the galaxy and state whether it is moving towards or away from Earth.

3 The red shift of dark lines in an absorption spectrum from a distant galaxy has $z = 0.064$. Calculate the speed of recession of the galaxy.

Cosmological red shift

The discovery that all the very distant galaxies were moving away from Earth *and* that the speed they were moving away depended on their distance away led to the theory that space was expanding rather than the galaxies moving through space. The galaxies closer to us, such as Andromeda, which has a blue shift, are moving through space relative to the Earth. For the very distant galaxies this effect is not as noticeable as the fact that they are all moving away due to the expansion of space. Although their red shift is calculated in the same way it is not referred to as a Doppler shift but as a cosmological red shift. The diagram illustrates the difference.

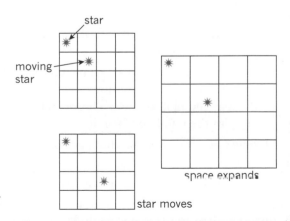

15.2 The expanding Universe

Units of distance

The light year

The light year is the distance travelled by light in 1 year.

The speed of light, c, is calculated in $\mathrm{m\,s^{-1}}$: $c = 3.00 \times 10^8\,\mathrm{m\,s^{-1}}$.

1 light year (ly) $= (3.00 \times 10^8\,\mathrm{m\,s^{-1}})(\text{seconds in a year}) \approx 9.46 \times 10^{15}\,\mathrm{m}$

The astronomical unit

The astronomical unit is approximately equal to the mean Earth–Sun distance. It is defined as:

1 astronomical unit (AU) $= 149\,597\,870\,700\,\mathrm{m} \approx 1.5 \times 10^{11}\,\mathrm{m}$

The parsec

The parsec is the distance from the Sun to an object with a parallax angle of 1 arcsecond.

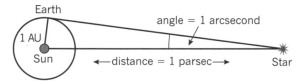

$$1 \text{ parsec (pc)} = \frac{1\,\mathrm{AU}}{\tan(1'')} = \frac{(1.496 \times 10^{11}\,\mathrm{m})}{\tan(1° \div 3600)} \approx 3.09 \times 10^{16}\,\mathrm{m}$$

WORKED EXAMPLE

Calculate the value of 1 parsec in light years.

$1\,\mathrm{ly} = 9.46 \times 10^{15}\,\mathrm{m}$ $1\,\mathrm{pc} = 3.09 \times 10^{16}\,\mathrm{m}$

$$1\,\mathrm{pc} = \frac{(3.09 \times 10^{16}\,\mathrm{m})}{(9.46 \times 10^{15}\,\mathrm{m\,ly^{-1}})} = 3.26\,\mathrm{ly}$$

PRACTICE QUESTIONS

1 Calculate the value of the light year from the speed of light and seconds in 1 year.

2 Rigel is 722.9 ly from Earth. Calculate this distance in parsecs.

Hubble's law

Hubble measured red shifts of galaxies and noticed these were greater for more distant galaxies. He plotted a graph of speed of recession (speed moving away) against distance to the galaxy.

The values cover such a wide range that logarithmic scales are needed. This is graph paper with logarithmic instead of linear spacing, as shown on the left-hand scale of the diagram on the next page. You could look up the logs and plot them on a linear graph, as shown on the right-hand side, but log graph paper allows you to plot the numbers directly.

Hubble's graph shows that the speed of recession of the galaxies, v, is proportional to their distance away, d.

$v = H_0 d$ where $H_0 =$ the Hubble constant **(1)**

The Hubble constant is the gradient of the graph of v against d, but notice that there is a large uncertainty in its value due to the difficulty of measuring the distances and red shifts of distant galaxies. The units are often given as $\mathrm{km\,s^{-1}\,Mpc^{-1}}$ as v is speed in $\mathrm{km\,s^{-1}}$ and d is distance in Mpc.

Logarithmic scale

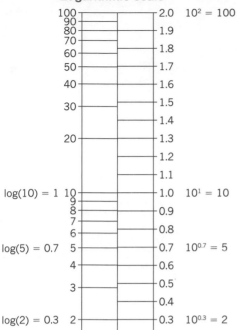

Hubble's law

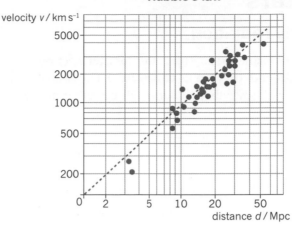

Some of the most important evidence for the expansion of the Universe comes from Hubble's law:

Distant galaxies are all moving away. The most distant galaxies are moving away faster and their speed is proportional to their distance away from us.

This suggests that space is expanding and the red shifts are cosmological red shifts (see Topic 15.1).

This supports the Big Bang theory because if they are all moving away then at some time in the past they were all at the same point. The distance travelled is given by:

$d = vt$ where t is the time since the Big Bang, the age of the Universe.

Using equation **(1)** $\dfrac{v}{H_0} = vt$

The age of the Universe, $t = \dfrac{1}{H_0}$

WORKED EXAMPLE

Calculate the age of the Universe in years assuming the Hubble constant $= 74.3\,\text{km s}^{-1}\,\text{Mpc}^{-1}$.

$$\text{Age of the Universe} = \frac{1}{74.3\,\text{km s}^{-1}\,\text{Mpc}^{-1}}$$

$$\text{Age of the Universe} = \frac{(10^6 \times 3.09 \times 10^{16}\,\text{m})}{(74.3 \times 10^3\,\text{m s}^{-1})} = 4.16 \times 10^{17}\,\text{s} = \frac{4.16 \times 10^{17}\,\text{s}}{(365.25 \times 24 \times 60 \times 60)\,\text{s year}^{-1}}$$

$$= 1.32 \times 10^{10}\,\text{years}$$

PRACTICE QUESTIONS

3 The uncertainty in the value of the Hubble constant given above is $\pm\,2.1\,\text{km s}^{-1}\,\text{Mpc}^{-1}$. Calculate an upper and lower limit for the age of the Universe.

4 Use the value of the Hubble constant above to calculate the distance to the most distant galaxy observed with a red shift $z = 8.6$.

15.3 Stars

Black body radiation

Black body is the name for an object that radiates electromagnetic radiation that depends only on its temperature. The diagram shows the shape of black body radiation curves for objects at different temperatures.

As the temperature, in kelvin, of the body increases, the wavelength of the peak radiation emitted, λ_{max}, decreases. This is called Wien's law:

$\lambda_{max}T = 2.89 \times 10^{-3}$ m K (note that these units are metre × kelvin)

The radiation flux F emitted from the surface of the body also increases with temperature T.

$F = \sigma T^4$ where the Stefan constant $\sigma = 5.67 \times 10^{-8}$ W m^{-2} K^{-4}

The power radiated from a surface of area A is:

$$P = \sigma A T^4$$

The power radiated from, or luminosity of a spherical star ($A = 4\pi r^2$) is $L = 4\pi r^2 \sigma T^4$.

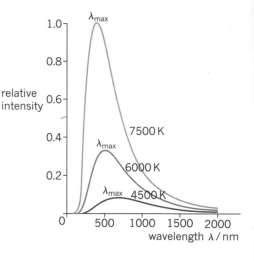

WORKED EXAMPLE

The star, Rigel, has a luminosity of about 4.5×10^{31} W and a surface temperature of 1.2×10^4 K. Calculate its radius. Determine the colour of Rigel.

Using the equation for luminosity:

$$r^2 = \frac{(4.5 \times 10^{31} \text{ W})}{4\pi(5.67 \times 10^{-8} \text{ W m}^{-2}\text{K}^{-4})(1.2 \times 10^4 \text{ K})^4} = 3.05 \times 10^{21} \text{ m}^2$$

$r = 5.5 \times 10^{10}$ m

Colour will depend on the peak wavelength:

$$\lambda_{max} = \frac{2.89 \times 10^{-3}}{T} \text{ m K} = \frac{2.89 \times 10^{-3} \text{ m K}}{(1.2 \times 10^4 \text{ K})} = 2.4 \times 10^{-7} \text{ m}$$

(This is the UV region so Rigel will appear blue-white.)

PRACTICE QUESTIONS

1 The temperature of the Sun is 5800 K. Calculate the peak wavelength of its emission spectrum.

2 A star has a surface temperature of 3500 K.
   a Determine its apparent colour.
   b It has a radius of 8.2×10^8 m. Calculate its luminosity.

STRETCH YOURSELF!

The peak wavelength in the black body radiation curve for Arcturus is 670 nm.
a Calculate the surface temperature.
b The luminosity is 170 times greater than the Sun. Calculate the radius of Arcturus.
   (Surface temperature of the Sun = 5800 K, radius of the Sun = 6.96×10^8 m.)

Hertzsprung–Russell diagram

This is a diagram of luminosity against decreasing temperature in kelvin.

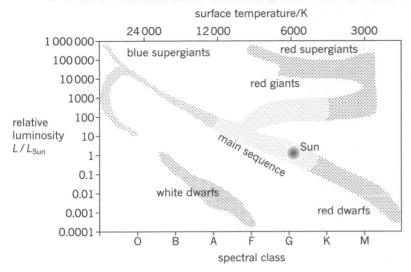

The luminosity is in units of the Sun's luminosity, L_{Sun}. The numbers are so large that logarithmic scales are used (see Topic 15.2). It seems strange to plot the temperature decreasing, but it corresponds to the wavelength increasing from blue to red. Notice that the black body curves above also have wavelength increasing to the right.

The spectral type is a classification system for stars based on colour.

WORKED EXAMPLE

Describe what the position of the Sun on the Hertzsprung–Russell diagram tells you about it. Describe what will happen as it finishes fusing hydrogen.

The Sun has a luminosity of 1 on this scale. It is a G-type star with a surface temperature of about 6000 K (5800 K). It is on the main sequence.

Stars spend most of their time on the main sequence. When the Sun finishes up its supply of hydrogen, its size will increase and temperature will decrease and it will become a red giant, moving up and to the right on the diagram, as it fuses helium. When the helium has been fused to carbon the Sun will collapse and heat up. As it gets smaller (and so less luminous) and hotter it will become a white dwarf, moving down and to the left. Then it will cool and fade, moving to the right and down.

PRACTICE QUESTION

3 Sketch the axes of a Hertzsprung–Russell diagram.
  a Divide the diagram into four quarters and label them:
      i hot and bright   ii hot and dim
    iii cool and bright   iv cool and dim.
  b Using information calculated and given on this page, mark on your diagram:
      i the Sun   ii Rigel   iii Arcturus.

1 An automatic switch is used with power supply $V = 6.0\,V$ so that a capacitor with capacitance C charges and discharges repeatedly through a resistor of $110\,\Omega$. The capacitor discharges $f = 350$ times per second and the current through it is $I = 9.9\,mA$.

Show that $C = \dfrac{I}{fV}$ and calculate the capacitance C.

2 The following data is for the discharge of a capacitor through a $1.6\,M\Omega$ resistor.

Plot a suitable graph to show the discharge is exponential, and determine the capacitance of the capacitor.

Time t/s	0.0	5.0	10.0	15.0	20.0	25.0	30.0	35.0	40
p.d. V/V	20.0	14.0	9.79	6.85	4.79	3.35	2.35	1.65	1.16

3 A radioactive isotope has a half-life of 88 years. When an atom decays it releases $5.6\,MeV$ of energy. Initially, a sample of the isotope has 2.4×10^{24} atoms.

Calculate the total energy released in 66 years. Give your answer in joules.

4 The following data was collected from a sample of radioactive material. The mean background count before the sample was tested was $3\,Bq$.

   a Copy the table and add rows to show the corrected count in C and $\ln C$.

   b Plot a suitable graph and determine the half-life of the material.

Time t/s	0	100	200	300	400	500	600
Count rate $/\times 10^3\,Bq$	123	87	73	49	39	31	23

5 A $7.50\,g$ sample of ancient bone has an activity of $6.2\,Bq$ due to the carbon-14 content. A $12.0\,g$ sample of modern bone has an activity of $6.5\,Bq$. Calculate the age of the ancient bone.

Half-life of carbon-14 $= 5730$ years.

6 The half-life of technetium-99m is 6.01 hours. Calculate:

   a the decay constant of technetium-99m

   b the number of atoms in a sample with activity $750\,MBq$

   c the time taken for the activity of a sample to fall from $750\,MBq$ to $500\,Bq$. Give your answer to the nearest hour.

7 **a** Complete the following equation for the fission of plutonium-239:

$$^{1}_{0}n + {}^{239}_{94}Pu \rightarrow {}^{134}_{54}Xe + {}^{103}\underline{\quad}Zr + \ldots {}^{1}_{0}n$$

   b Calculate the mass defect in this reaction.

   c Calculate the binding energy released in MeV.

Mass: neutron = 1.0087 u

plutonium-239 = 239.0521 u

xenon-134 = 133.9054 u

zirconium-103 = 102.9266 u

8 Use the graph below to answer these questions:

a Identify the least stable three nuclides and justify your answer.

b Identify the most stable three nuclides and justify your answer.

c Calculate the change in total binding energy when two hydrogen-1 nuclei fuse to give a helium-4 nucleus. State whether this binding energy is released or must be supplied.

d Explain how the graph shows that both fission and fusion can release energy.

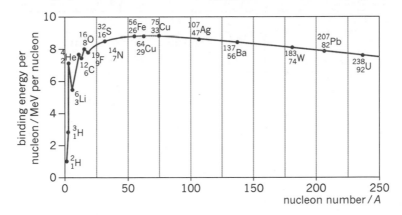

9 On a clear cloudless night the Earth's surface radiates heat. The temperature of the ground is 8 °C.

Calculate the rate of heat loss per metre squared of the Earth's surface, assuming it radiates as a black body.

10 The star Capella is 45.5 light years from Earth and the radiation flux reaching Earth from Capella is $1.2 \times 10^{-8} \, \text{W m}^{-2}$. The surface temperature is 4940 K.

a Calculate the distance to the star in metres.

b Assuming the star radiates like a black body, calculate its radius.

DATA AND FORMULAE

Constants and AS formulae

Constants

These values are used in this book and are typical of those given on A Level data and formulae sheets.

Always use the value you are given or you may lose marks. Do not give your answer to more significant figures than are given in the constant because if the constant is not accurate to that many figures then your answer may not be accurate.

Constant or value	Symbol	Value
speed of light in a vacuum	c	$3.00 \times 10^8 \, \text{m s}^{-1}$
permittivity of free space	ε_0	$8.85 \times 10^{-12} \, \text{C}^2 \, \text{N}^{-1} \, \text{m}^{-2}$
magnitude of charge on the electron	e	$1.60 \times 10^{-19} \, \text{C}$
Planck constant	h	$6.63 \times 10^{-34} \, \text{J s}$
gravitational constant	G	$6.67 \times 10^{-11} \, \text{N m}^2 \, \text{kg}^{-2}$
Avogadro constant	N_A	$6.02 \times 10^{23} \, \text{mol}^{-1}$
molar gas constant	R	$8.3 \, \text{J K}^{-1} \, \text{mol}^{-1}$
Boltzmann constant	k	$1.38 \times 10^{-23} \, \text{J K}^{-1}$
Stefan constant	σ	$5.67 \times 10^{-8} \, \text{W m}^{-2} \, \text{K}^{-4}$
electron rest mass	m_e	$9.11 \times 10^{-31} \, \text{kg}$
proton rest mass	m_p	$1.68 \times 10^{-27} \, \text{kg}$
neutron rest mass	m_n	$1.67 \times 10^{-27} \, \text{kg}$
acceleration of free fall	g	$9.81 \, \text{m s}^{-2}$
gravitational field strength close to Earth	g	$9.81 \, \text{N kg}^{-1}$
unified atomic mass unit	u	$1.661 \times 10^{-27} \, \text{kg}$

AS formulae

This is a list of the formulae used in this book for AS.

Your A Level specification will have a list of formulae for AS that you will be given in the exam. This list is typical of the equations that may be on it.

It is a good idea to check this list against your specification formulae list and delete those not on it. If they are not on your specification at all you can ignore them, but if they are formulae you need to know and you are not given them in the exam, highlight them and learn them.

Waves

Wave speed $\quad v = f\lambda$

Period and frequency $\quad f = \dfrac{1}{T}$

Fringe spacing $\quad x = \dfrac{\lambda D}{a}$

Or Fringe spacing $\quad w = \dfrac{\lambda D}{s}$

Diffraction grating $\quad d \sin \theta = n\lambda$

Refractive index $\quad n_s = \dfrac{c}{c_s}$

Refractive index $\quad _1 n_2 = \dfrac{n_2}{n_1}$

Law of refraction $\quad n_1 \sin \theta_1 = n_2 \sin \theta_2$

Critical angle where $n_1 > n_2$ $\quad \sin \theta_c = \dfrac{n_2}{n_1}$

Mechanics

Velocity $\quad v = \dfrac{\Delta x}{\Delta t}$

Acceleration $\quad a = \dfrac{\Delta v}{\Delta t}$

Equations of motion:

$s = \frac{1}{2}(u + v)t$

$v = u + at$

$s = ut + \frac{1}{2}at^2$

$v^2 = u^2 + 2as$

Resolving forces $\quad F_x = F\cos\theta$

$F_y = F\sin\theta$

Force $\quad F = ma$

Weight $\quad W = mg$

Work done $\quad W = Fs\cos\theta$

Kinetic energy $\quad E_k = \frac{1}{2}mv^2$

Potential energy $\quad \Delta E_p = mg\Delta h$

Power $\quad P = \dfrac{\Delta W}{\Delta t}$

Power $\quad P = Fv$

Efficiency $= \dfrac{\text{useful power output}}{\text{total power input}} \times 100\%$

Materials

Density $\quad \rho = \dfrac{m}{V}$

Pressure $\quad p = \dfrac{F}{A}$

Hooke's law $\quad F = kx$

Or Hooke's law $\quad F = k\Delta L$

Energy stored $\quad E = \frac{1}{2}Fx$ or $E = \frac{1}{2}F\Delta L$

Energy stored $\quad E = \frac{1}{2}kx^2$ or $E = \frac{1}{2}k\Delta L^2$

Tensile stress $\quad \sigma = \dfrac{F}{A}$

Tensile strain $\quad \varepsilon = \dfrac{x}{L}$ or $\varepsilon = \dfrac{\Delta L}{L}$

Young modulus $\quad E = \dfrac{\text{tensile stress}}{\text{tensile strain}}$

Electricity

Resistance $\quad R = \dfrac{V}{I}$

Total resistance $\quad R_T = R_1 + R_2 + R_3 + \ldots$

Total resistance $\quad \dfrac{1}{R_T} = \dfrac{1}{R_1} + \dfrac{1}{R_2} + \dfrac{1}{R_3} + \ldots$

Resistivity $\quad \rho = \dfrac{RA}{L}$

Charge $\quad \Delta Q = I\Delta t$

Current $\quad I = nAve$

Energy $\quad W = QV$

Power $\quad P = IV = I^2R = \dfrac{V^2}{R}$

emf $\quad \varepsilon = \dfrac{E}{Q}$

emf $\quad \varepsilon = I(R + r)$

Potential divider $\quad V_{out} = \dfrac{R_2}{(R_1 + R_2)}V_{in}$

Quantum physics

Photon energy $\quad E = hf$

Photon energy $\quad E = \dfrac{hc}{\lambda}$

Photoelectric effect $\quad hf = \phi + \frac{1}{2}m_e v_{max}^2$

De Broglie wavelength $\quad \lambda = \dfrac{h}{p} = \dfrac{h}{mv}$

A2 Formulae

This is a list of the formulae used in this book for A2.

Your A Level specification will have a list of formulae for A2 that you will be given in the exam. This list is typical of the equations that may be on it.

It is a good idea to check this list against your specification formulae list and delete those not on it. If they are not on your specification at all you can ignore them, but if they are formulae you need to know and you are not given them in the exam, highlight them and learn them.

More forces and motion

Force $\quad F = \dfrac{\Delta(mv)}{\Delta t}$

Force $\quad F = \dfrac{\Delta p}{\Delta t}$

Impulse $\quad F\Delta t = \Delta(mv)$

Angular velocity $\quad \omega = \dfrac{v}{r}$

$$\omega = 2\pi f$$

$$\omega = \dfrac{2\pi}{T}$$

Centripetal acceleration $\quad a = \dfrac{v^2}{r} = \omega^2 r$

Centripetal force $\quad F = \dfrac{mv^2}{r}$

Centripetal force $\quad F = m\omega^2 r$

Oscillations

Acceleration $\quad a = -(2\pi f)^2 x$

Maximum acceleration $\quad a = -(2\pi f)^2 A$

Speed $\quad v = \pm(2\pi f)\sqrt{(A^2 - x^2)}$

Maximum speed $\quad v = \pm(2\pi f)A$

Displacement $\quad x = A\cos(2\pi f)t$

Pendulum $\quad T = 2\pi\sqrt{\dfrac{l}{g}}$

Mass on a spring $\quad T = 2\pi\sqrt{\dfrac{m}{k}}$

Thermal physics

Gas law $\quad pV = nRT$

Gas law $\quad pV = NkT$

Kinetic theory $\quad pV = \frac{1}{3}Nmc^2_{\text{rms}}$

KE of gas molecule $\quad E_k = \frac{1}{2}mc^2_{\text{rms}}$

KE of gas molecule $\quad E_k = \dfrac{3kT}{2}$

Fields

Gravitational force $\quad F = -\dfrac{GMm}{r^2}$

Gravitational field strength $\quad g = \dfrac{F}{m}$

Gravitational field strength $\quad g = -\dfrac{GM}{r^2}$

$$V = \dfrac{-GM}{r}$$

Orbital motion $\quad T^2 = \dfrac{4\pi^2 r^3}{GM}$

Electric force $\quad F = \dfrac{Qq}{4\pi\varepsilon_0 r^2}$

Electric field strength $\quad E = \dfrac{F}{Q}$

Radial field $\quad E = \dfrac{Q}{4\pi\varepsilon_0 r^2}$

Electric potential $\quad V = \dfrac{Q}{4\pi\varepsilon_0 r}$

Uniform field $\quad E = \dfrac{V}{d}$

Magnetic force $\quad F = BIl\sin\theta$

Magnetic force $\quad F = BQv$

Magnetic flux $\quad \phi = BA\cos\theta$

Magnetic flux linkage $\quad N\phi = BAN\cos\theta$

Induced emf $\quad \varepsilon = -\dfrac{N\Delta\phi}{\Delta t}$

Capacitors

Capacitance $C = \dfrac{Q}{V}$

Total capacitance $\dfrac{1}{C} = \dfrac{1}{C_1} + \dfrac{1}{C_2} + \dots$

Total capacitance $C = C_1 + C_2 + \dots$

Energy stored $W = \frac{1}{2}QV$

Energy stored $W = \frac{1}{2}CV^2$

Charge on capacitor $Q = Q_0 e^{-t/RC}$

Time constant $\tau = RC$

Radioactivity

Radioactive decay $\dfrac{dN}{dt} = -\lambda N$

Radioactive decay $N = N_0 e^{-\lambda t}$

Decay constant $\lambda = \dfrac{\ln(2)}{T_{\frac{1}{2}}}$

Activity $A = -\lambda N$

Activity $A = A_0 e^{-\lambda t}$

Energy, mass $E = mc^2$

Energy, mass $\triangle E = \triangle mc^2$

Medical physics

Intensity $I = \dfrac{k}{r^2}$

Intensity radiated from point source $I = \dfrac{P}{4\pi r^2}$

Attenuation $I = I_0 e^{-\mu x}$

Astrophysics

Doppler shift when $v >> c$ $z = \dfrac{\triangle f}{f} = -\dfrac{\triangle \lambda}{\lambda} = \dfrac{v}{c}$

Hubble's law $v = H_0 d$

Age of the Universe $t = \dfrac{1}{H_0}$

Wien's law $\lambda_{max} T = 2.89 \times 10^{-3}\,\text{m K}$

Stefan's law $P = \sigma A T^4$

Index

A

A2 formulae **126–7**
acceleration **24**
    centripetal force and acceleration **65**
activity **104–5**
air resistance **38**
alpha decay **106**
angular velocity **64–5**
antimatter **112–13**
AS formulae **124–5**
astronomical units **118**
astrophysics **116–17**
    expanding Universe **118–19**
    red shift **116–17**
    stars **120–1**
atomic mass unit **107**

B

base units **6–7**
    beta decay **106**
binding energy **108**
    binding energy per nucleon **108**
black body radiation **120**
Boltzmann constant **77**
Boyle's law **72**

C

capacitors **96–7**
centripetal force **84**
    centripetal force and acceleration **65**
change of phase **78**
Charles's law **72–3**
circuits **53**
    capacitors in circuits **96–7**
    RC circuits **98–101**
    solving circuit problems **57**
circular motion **64–5**
    angular velocity **64–5**
    centripetal force and acceleration **65**
    radians **64**
collisions **43**
    elastic and inelastic collisions **44–5**
compression **56**
conservation of momentum **42–3**
constants **124**
    Boltzmann constant **77**
    time constant **98–9**
    universal gas constant **74–5**
    using base units to check constants **6–7**
converging lens **22**
Coulomb's law **86**
critical angle **21**
curves **27**

D

data **124–7**
de Broglie equation **61**
density **40–1**
derived units **6**
diffraction **18–19**
    diffraction grating **19**
    Young's slits **18**
dimensions **6–7**
displacement–time graphs **14–15, 24–5**
distance travelled **26–7**
    counting squares **27**
    curves **27**
    trapezium **26**
    triangles and rectangles **26**
Doppler effect **116–17**

E

$E = mc^2$ **107**
efficiency **37**
elastic collisions **44–5**
elasticity **46–7**
    elastic potential energy **47**
    Hooke's law **46–7**
    stress and strain **48–9**
electric charge **52**
    mean drift velocity **52–3**
electric currents **53**
electric fields **86–7**
    Coulomb's law **86**
    parallel plates **87**
    point charge **86–7**
electric potential **88–9**
electricity **50–1**
    current in circuits **53**
    electric charge **52–3**
    potential divider and other circuits **56–7**
    resistance and resistivity **50–1**
electromagnetic induction **92–3**
    induced emf **92–3**
    magnetic flux **92**
emf (electromotive force) **54–5**
    energy, power and pd **54**
    induced emf **92–3**
energy **36–7, 54**
    binding energy **108**
    elastic potential energy **47**

electric potential energy **88**
energy levels **60–1**
energy of a photon **58**
energy stored by a capacitor **97**
energy transfer **68–9**
internal energy **78–9**
equations
de Broglie equation **61**
equation of state for an ideal gas **74**
kinematic equations **28–9**
kinetic theory equation **76**
nuclear equations **106–7**
using base units to check equations **6–7**
wave equation **15**
explosions **43**
exponential change **98**
exponential decay **98**
exponential discharge **98**

F
fission **108–9**
binding energy **108**
forces **30–1**
centripetal force **84**
centripetal force and acceleration **65**
force on a moving charge **90–1**
perpendicular forces **32–3**
resultant forces **30–1**
formulae **124–7**
free-fall **38**
fundamental particles **110–13**
leptons **110**
matter and antimatter **112–13**
mesons **112**
quarks **111**
standard model **110**
fusion **108–9**
binding energy **108**

G
gases
gas laws **72–3**
ideal gas **74–5**
graphs
displacement–time graphs **14–15, 24–5**
graphs of SHM **68**
graphs of T and T2 against l for a
pendulum **12–13**
log graphs **100–1**
pendulums **12–13**
resistance from a graph **50–1**

terminal velocity sketch graph **39**
velocity–time graphs **25**
waves **14–15**
gravity **38–9**
gravitational field **80–1, 82**
gravitational force **80**
gravitational potential **82–3**

H
half-life **103, 104**
Hertzsprung–Russell diagram **121**
Hooke's law **46–7**
Hubble's law **118–19**

I
ideal gas **74–5**
equation of state for an ideal gas **74**
universal gas constant **74–5**
impulse **42**
inelastic collisions **44–5**
interference **17, 18–19**
interference patterns **18**
Young's slits **18**
internal energy **78–9**
change of phase **78**
ideal and real internal energy **78–9**
specific heat capacity **78–9**
specific latent heat **78**

K
kelvin scale **72**
kinematic equations **28–9**
projectiles **29**
kinetic theory **76–7**
Boltzmann constant **77**

L
law of refraction **20–1**
lenses **22**
leptons **110**
light years **118**
logarithms **84–5**
logs **100**
log graphs **100–1**
longitudinal waves **14**

M
magnetic fields **90–1**
force on a moving charge **90–1**
magnetic force and flux density **90**
magnetic flux **92**

mass on a spring **71**
materials **46–9**
matter and antimatter **112**
    pair production **112–13**
measurement uncertainties **8**
medical physics **114–15**
mesons **112**
momentum **42–3**
    conservation of momentum **42–3**
    impulse and momentum **42**
    momentum and energy **44–5**
motion **24–5**
    circular motion **64–5**
    displacement–time graphs **24–5**
    distance travelled **26–7**
    kinematic equations **28–9**
    orbital motion **84–5**
    SHM (simple harmonic motion) **66–71**
    speed, velocity and acceleration **24**
    velocity–time graphs **25**
    vertical motion and gravity **38–9**

N
Newton's laws **34–5**
nuclear equations **106–7**
    alpha decay **106**
    atomic mass unit **107**
    beta decay **106**
    $E = mc^2$ **107**
nuclear fission **108–9**
nuclear fusion **108–9**
nuclear physics **106–9**
nuclear reactions **106–7**
nucleons **108**

O
orbital motion **84–5**
    centripetal force **84**
    power laws **84–5**
order of magnitude calculations **10, 11**

P
parsecs **118**
particle physics **110–13**
    particle theory **11**
    wave particle duality **61**
pd (potential difference) **54**
pendulums **12–13, 70–1**
    graphs of T and T^2 against l for a
        pendulum **12–13**

    testing a relationship **12**
    using the graph **13**
perpendicular forces **32–3**
photoelectric effect **58–9**
    energy of a photon **58**
photons **58**
potential divider **56–7**
power **37, 54**
    power laws **84–5**
powers of ten **11**
prefixes **10, 11**
pressure **41**
    pressure law **73**
projectiles **29**

Q
quantum physics **58–61**
    energy, waves and particles **60–1**
    photoelectric effect **58–9**
quarks **111**

R
radians **64**
radioactive decay **102–3, 104**
    exponential decay **102–3**
    half-life **103, 104**
radioactivity **102–3**
    activity **104–5**
RC circuits **98–101**
    exponential change **98**
    exponential decay **98**
    log graphs **100–1**
    natural logs **100**
    time constant **98–9**
rectangles **26**
red shift **116–17**
    cosmological red shift **117**
    Doppler effect **116–17**
    red and blue shifts **117**
refraction of light **20–1**
    critical angle **21**
    law of refraction **20–1**
    refractive index **20**
    total internal reflection **21**
resistance **50**
    emf and internal resistance **54–5**
resistivity **51**
resolving forces **32–3**
resultant forces **30–1**

S

scalars **24**
SHM (simple harmonic motion) **66–7**
    energy transfer **68–9**
    graphs of SHM **68**
    mass on a spring **71**
    pendulums **70–1**
significant figures **9**
    calculations using uncertainties and significant
        figures **9**
    how many significant figures to use? 9
    number of significant figures **9**
small numbers **11**
specific heat capacity **78–9**
specific latent heat **78**
speed **24**
squares **27**
standard form **10, 11**
stars **120–1**
    black body radiation **120**
    Hertzsprung–Russell diagram **121**
stationary waves **23**
    open and closed pipes **23**
    stretched string **23**
strain **48–9**
stress **48–9**
superposition of waves **16–17**
    phase difference **16**
    principle of superposition of waves **17**

T

temperature scales **72**
tension **46**
terminal velocity **38–9**
thermal physics **72–3**
    gas laws **72–3**
    ideal gas **74–5**
    internal energy **78–9**
    kelvin scale **72**
    kinetic theory **76–7**
time constant **98–9**
total internal reflection **21**
transverse waves **14**
trapeziums **27**
triangles **26**

U

uncertainties **8–9**
    calculations using uncertainties and significant
        figures **9**
    examples of uncertainties in measurements **8**
units **6–7**
    astronomical units **118**
    base units **6–7**
    derived units **6**
Universe, expanding **118–19**
    Hubble's law **118–19**
    units of distance **118**
Universe, scale of **10–11**
    order of magnitude calculations **10, 11**
    powers of ten **11**
    prefixes **10, 11**
    using standard form **10, 11**
upthrust **41**

V

vectors **24**
velocity **24**
    angular velocity **64–5**
    mean drift velocity **52–3**
    terminal velocity **38–9**
vertical motion **38–9**

W

waves **14–15**
    displacement–time graphs **14–15**
    phase difference **16**
    stationary waves **23**
    superposition of waves **16–17**
    transverse and longitudinal waves **14**
    wave equation **15**
    wave particle duality **61**
work **36–7**
    work done in a gravitational field **82**

X
X-rays **114–15**

Y
Young's slits **18**

Notes

Notes

Notes